HIGHRISE Horticulture

A Guide to Gardening in Small Spaces

David Tarrant

Whitecap Books
Vancouver • Toronto

Whitecap Books
Vancouver/Toronto

Edited by Elaine Jones
Illustrations by Peter Lynde and Brad Nickason
Cover Illustration by Brad Nickason
Design by Brad Nickason
Page Layout and Assembly by Opus Productions Inc.

Printed and bound in Canada by Friesen Printers, Altona, Manitoba

Canadian Cataloguing in Publication Data

Tarrant, David.
Highrise horticulture

ISBN 0-921061-49-8
1. Patio gardening. 2. Indoor gardening
I. Title.
SB419.T37 1989 635.9'671 C89-091404-4

To all my friends.

Table of Contents

Acknowledgments

I would like to thank all the students who have attended our yearly workshops on small space and patio gardening at the UBC Botanical Garden extension program. Many of them invited me to see their tiny gardens and shared their growing techniques with me.

Introduction

Welcome to the small-space gardening book. Whether you want to decorate an apartment, townhouse or condominium, or even your office desk, this book is written with you in mind. Small spaces can be turned into blossoming and sometimes fruitful gardens. All that is required is enthusiasm, some thought and a little hard work.

Over a period of fifteen years I have had several apartment gardens, and probably one of the nicest things about them is that after puttering among your planters for an hour or less, you can sit back and enjoy without having to face endless chores of weeding or mowing. It allows you to experience all the fun of gardening with time left over to pursue other interests and hobbies.

In this book I share ideas and experiences that have worked for me, most of which are suitable for all areas of the country. A lot of the most useful tips have been passed on to me by the many new friends I have made while teaching workshops and classes on this very subject.

While this book is written primarily for Canada, those of you living in the northern areas of the United States will find the information applicable to your area also. The only difference is that when you look for the hardier permanent plants for your balcony or patio, you will have to check with your local garden centers and agricultural extension agents in your region. Better yet, experiment on your own!

I hope that when you have read this book, it will inspire you to do your part in the greening of the concrete jungle, and that you will have at least a couple of pots on your patio soon.

I

Flowering Balconies and Patios

Before planning a balcony garden, check with the manager or owner to see if balcony plants are allowed. I hope everyone can have at least a couple of pots on the balcony or patio.

Very often balconies are plagued by wind, as they are perched on the side of manmade cliffs. If the balcony is the open rail type, some sort of windbreak is beneficial. The easiest solution is to install clear fiberglas or similar heavy plastic sheets. But if your apartment manager does not allow this, I would suggest either some permanent little evergreens for the front or scarlet runner beans woven in and out of the railings—an excellent summer screen. (For a more comprehensive list of hardy plant material, see the section on hardy outdoor plants in this chapter.)

Whenever any pots, boxes or planters are put on a

balcony they must have drainage holes, otherwise water builds up during wet seasons. Trays should be set underneath them to catch excess drips, just in case the person below is sunbathing while you are watering. Garden centers and plant shops sell a large assortment of drainage trays suitable for all pot and container sizes.

The type of containers used for planting can be anything from tin cans and plastic bags to $500.00 ceramic planters. Plants are not fussy and will grow in just about anything. There is a lady in New York City who landscapes balconies using large plastic garbage bags as planters. She fills them with soil and plants them up, arranging them around the edges of the balconies like a flower border. Then she places manmade rock around the base to hide the bags.

Homemade wooden planters are acceptable, but do remember to treat the wood first with a preservative inside and out. Make sure you do it during the winter months of January and February so that it has time to be absorbed into the wood and fully dried. Wood preservative in its liquid form is toxic to plants. It's a good idea to line a wooden planter with plastic, making sure to put holes in the bottom before filling the planter with soil. If clay pots are used, they should be soaked an hour or two in water before planting. Dry clay is absorbent, and will leach moisture out of the soil if the pots are not thoroughly moistened.

The most important factor is the size of the container. It should be at least 30 x 30 cm (12 x 12 inches), especially if you are not around to water all day. Small containers dry out far too quickly once they are full of roots. The pots or wooden containers can be set at varying levels so that the plant from one hides the pot of the other. Also, this way you get more plants per area. Driftwood and rocks in between can give a natural effect.

Climbing plants will need support, such as chicken wire on the front of the balcony. Many garden centers sell rustic trellis which can be anchored by bolting onto the back of a wooden planter. Pieces of fishing net or garden net can be hung from the balcony above. (This is also a good way to get to know your neighbor.)

The type of soil used in containers is a problem because we all have a relative or friend in the country who is willing to give us some dark, healthy looking soil. Let me clarify this myth about dark soil. Color means nothing. It could be red or beige. The nutrients in the soil, not the color, are important for the growing of plants. Hydroponics proves that for the doubter. Please use sterilized soil for your balcony or patio garden, as it eliminates the introduction of weed seeds and many pests and diseases. If you are the adventurous type and would like to sterilize small amounts of soil in the oven, you can fill roasting bags with moist soil, seal them, then pop them into the oven until the temperature of the center of the soil has reached 82°C (178°F) as indicated on your meat thermometer.

Many people use prepackaged potting mixes which consist of vermiculite and peat with added nutrients and all the necessary trace elements. These latter elements are still giving problems to the commercial people, but by all means experiment for yourself. On my balcony I use this combination, well mixed: six parts by bulk sterilized soil, three parts by bulk moist peat, two parts by bulk coarse sand. (Do not use beach sand from the coast, as the salt content is too high.)

To each 36 liters (1 bushel) of this mix, which fills approximately three 12-liter (3-gallon) household buckets, I add 170 grams (6 ounces) of a slow-release fertilizer which releases its ingredients over a given period of time stated on the package. However, its lasting quality will be shorter in a container, as frequent watering leaches it out. From mid-July to September, weekly applications of liquid fertilizer are necessary. Five ml (1 teaspoon) of Hi-Sol 20-20-20 per each 4 liters (1 gallon) of water works well. Often this is more than you need, but it keeps well.

I realize that it is not practical for everyone to mix soil. Sterilized potting mix from your local garden center or nursery is fine. Ask what fertilizer has been added. If none has been added, use your own.

Before filling your planter, make sure it has adequate drainage holes in the bottom. If it is flat-bottomed and is to be placed on a flat surface, stand it on 2.5-cm (1-inch) blocks to allow passage of air underneath. This also allows the water to escape into the tray.

When filling, always put a layer of drainage in the bottom: broken pots, rocks, recycled styrofoam cups or crumpled plastic packs left over from when you purchased your bedding plants. The thickness of this layer will vary with container size, but it should just cover the bottom to stop the potting mix from washing out through the drainage holes. Never fill a pot or planter to the top; always leave a few centimeters (1/2 to 1 inch) between the surface of the soil and the rim of the container for watering purposes later on.

If you already have filled planters, each spring and fall add a little extra moist peat or fresh potting mix, with some balanced fertilizer added at 170 grams (6 ounces) per 36-liter (1-bushel) batch. Wherever possible, soil should be replaced every third year for best results.

Planting

With a few exceptions, summer annuals are basically the same across Canada. As some growing seasons are much shorter than others, I have made a list of approximate first and last frost dates for each province and territory. With balconies that face south, often the planting date can be earlier.

	Last Frost Date	First Frost Date
Newfoundland	first week June	last week September
Nova Scotia	mid-May	mid-October
New Brunswick	last week May	last week September
Prince Edward Island	mid-May	mid-October
Quebec	second week May	first week October
Ontario	third week May	last week October
Manitoba	third week May	mid-September
Saskatchewan	last week May	mid-September
Alberta	last week May	second week September
British Columbia		
(coastal region)	first week April	mid-November
(interior region)	first week May	first week October
Northwest Territories	last week June	last week August
Yukon	first week June	last week August

When your space is restricted, I think variety is the secret. We are all familiar with the traditional idea of a *Cordyline australis* (dracaena palm) in the center, pelargoniums around it and clumps of lobelia or alyssum at the edge. There is nothing wrong with this arrangement, except that it can be seen in any public area of any downtown.

My first planter was a half-barrel about 1 meter (3 feet) in diameter that contained one permanent evergreen (a mugho pine), runner beans at the back, one pelargonium, a miniature dahlia, one impatiens, one clump of chives and one of parsley, and a little lobelia squeezed over the front of the barrel. I admit that was a lot for one container, but with feeding, all went well, except that the evergreen was suffocated by the annual plants.

Since then, I have changed the plants and reduced the number per container. I grow beans separately. Do try a mixed planter, since it gives so much pleasure. When you are making a salad, it is nice to be able to pop out onto the balcony to gather fresh chives or herbs.

If you are worried about a permanent plant taking over the whole container, set out with the idea that it is only going to be with you for three years. Then donate it to a friend with a garden. The initial cost need not be great for a permanent plant. Watch for $1.49 specials at large department store garden shops or garden centers. If you keep it for three years, it works out to a cost of 50 cents for each year of pleasure; that is quite a bargain these days.

The type of permanent plant will vary according to the area of the country in which you live, but it is nice to have something living to look at in the winter months, even if it is only a little cedar. I know evergreens look rather dark when the temperature drops below zero, but they still provide visual relief from staring at cement walls all winter long.

If you are planning a summer show on a balcony that faces north, note that the following plants do well in the shade: tuberous begonia, common impatiens, schizanthus, fuchsia, sweet pea, coleus, monkey flower, nasturtium. These are good old faithfuls. When I had a balcony facing north, I found that marigolds, pelargoniums and quite a few others also did well, as long as they were mature flowering plants when planted. Once I saw a fascinating north-facing balcony with outdoor ferns and ivies planted among carefully placed driftwood and interesting rocks. The hardiness of these plants would have to be taken into consideration for the winter months.

On the south and west sides of the building, just about anything can be grown—although sometimes too much sun creates a problem. For very hot areas the following are sunny successes: Livingstone daisy, petunia, gazania, calendula, zinnia, dahlia, portulaca, lobelia, pelargonium, alyssum. Whichever way you face, bear in mind that one or two plants of each type are fine. You don't need to purchase a whole flat of each; a six-pack or individual pots of each will do nicely.

I'm assuming that you buy your plants rather than try to grow them from seed. Temperature is always the problem when raising seedlings, since so many plants, such as lobelia and petunia, need to be sown early in the year. The 22°C (73°F) temperature of an apartment is good for actual germination, but as soon as the seedlings are up, 12 to 16°C (52 to 61°F) is necessary for good growth; if it is too hot they damp off. Not many people are prepared to turn the heat down this low, so you can see that it is easier to buy your plants.

If you really enjoy sowing seeds, stick to the annuals that can be sown straight outside when the weather warms up. Some examples are calendula, bachelor's buttons, nasturtium, Shirley poppy and California poppy. These should be sown thinly. When they germinate, thin them out until they are 15 cm (6 inches) apart.

Unfortunately, it is human nature to buy bedding plants on the first warm weekend of the season and plant them straight outside. This can be a tremendous shock to bedding plants, especially as the temperature may drop close to freezing on the first night. Always introduce your plants to the outdoors gradually, by taking them in for the first two nights and putting them out during the day. For the next three nights, put a cardboard carton over them on the balcony. After a week of gradual introduction, your plants are ready to live outside.

When taking plants out of a flat or pot, always keep as many roots and as much soil on them as possible. Dig the hole a little larger and deeper than the roots are. After planting, firm the area with your fingertips so the roots cannot be pulled out easily. Then water them in.

Bedding plants inevitably have one pretty little flower on the top. Marigolds are a fine example. Please pick out the flower on each plant once they are planted on the balcony. People think I'm a murderer when I suggest this, but if you don't follow my advice, that plant will concentrate on making the one flower go to seed and it will forget about growing. If the top is pinched out, a stronger plant will be produced with a flower on every new shoot. Some nurseries pinch their plants at an early stage, so check when buying as there is no need to do it twice. On mature bedding plants, always pick out deadheads to ensure the production of more flowers throughout the summer.

If you live in a townhouse, you have an advantage over highrises because your ground-level dwelling has no wind problem or drip nuisance. Many townhouses have two garden areas, one of which is usually quite shaded. There are many plants that do well in the shade, particularly green shrubby plants. The other patio is often in full sun and all kinds of flowering plants and vegetables can be grown.

If there is soil in your townhouse garden it has probably been completely ruined during the construction period. You will have to make raised beds using 2.5-centimeter by 30-centimeter (1-inch by 12-inch) boards that have been treated with preservative. Fill these with good mix and proceed to plant. The same mixture that is used for containers would be ideal. Even small flowering shrubs can be used in this type of location. For variety and year-round color, stick to the mixed planting idea, as recommended for balconies.

Usually townhouses have a little area at the back door which is used for storing garbage cans. To hide such eyesores, you can make very pleasing screens, such as a trellis at the back with ivy or clematis climbing on it. Of course, your choice of greenery depends on local climate. If you want to use annuals, try trailing nasturtiums, morning glory, *Cobea scandens,* or sweet peas, which are usually successful. The advice for balcony gardens also applies to container growing outside townhouses.

I am often asked about large plants that can be kept out during summer and grown indoors during winter. This is not easy because any plant that has spent a warm summer outside starts to slow down in the fall. It does not necessarily lose leaves, but it rests. Suddenly it is lifted inside the apartment, which is 22°C (73°F) or thereabouts, with low humidity. For the plant, that is a summer temperature, but the nice strong sunlight is missing, so be sure to keep it by a well-lighted window. Some large plants that can spend the winter inside are *Fatsia japonica, Fatshedera Lizei,* ivy, *Pittosporum tobira,* bougainvillea, hibiscus, flowering maple, and *Tibouchina semidecandra.* These are not recommended for extreme northern areas.

Hardy Outdoor Plants

Here is a list of small ornamental shrubs to act as windbreaks on balconies. Planted in 12- or 20-liter (3- or 5-gallon) containers, they should last for three years, after which they are usually so big they have to be replaced. Those of us living in the more temperate regions of Canada can grow most of the plants on this list. I have included the hardiness zone numbers, which you can check against the map that is reproduced on succeeding pages.

Zone		
1	*Pinus mugo mughus*	mugho pine
1	*Potentilla fruticosa*	shrubby cinquefoil
1	*Shepherdia argentea*	silver buffaloberry
1B	*Cornus stolonifera*	red osier dogwood
1	*Lonicera (many species)*	honeysuckle
2	*Caragana aurantiaca*	pygmy peabush
2	*Crataegus chrysocarpa*	fireberry
2	*Juniperus communis saxitilis*	mountain juniper
2	*Juniperus communis*	common juniper
2	*Symphoricarpus albus*	snowberry
2B	*Picea abies 'nudiformis'*	bird's nest spruce
3	*Pachysandra terminalis*	pachysandra
3	*Taxus baccata repandens*	yew
3	*Thuja occidentalis*	American arborvitae
3	*Thuja orientalis*	Berckmann's Gold
5	*Buxus microphylla*	little leaf box
5	*Buxus microphylla koreana*	Korean box
5	*Juniperus chinensis*	Chinese juniper
5	*Mahonia aquifolium*	Oregon grape
6	*Pyracantha coccinea*	scarlet firethorn
7	*Prunus laurocerasus 'zabeliana'*	narrow-leaved cherry laurel
7B	*Viburnum davidii*	David viburnum
8B	*Fatsia japonica*	fatsia

The following perennial plants are so hardy that they grow well anywhere in Canada, even in extreme northern regions where the long summer days seem to make them work overtime.

Dicentra spectablilis	bleeding heart
Aquilegia	columbine
Delphinium elatum	delphinium
Lilium philadelphicum	western wood lily
Lupinus argenteus	lupin
Lychnis chalcedonica	Maltese cross
Aconitum napellus	monkshood

Phlox paniculata	phlox
Papaver nudicaule	Iceland poppy
Papaver orientale	oriental poppy
Rudbeckia hirta	brown-eyed susan
Veronica spicata	speedwell

The Hardiness Zones in Canada

The following map shows the areas of winter hardiness for ornamental plants in the more heavily populated areas of Canada. The map is based on a formula that takes into consideration several meteorological factors affecting the hardiness of a plant in a given location. The most important element in plant survival is the minimum temperature during the winter. Other important considerations are the length of the frost-free period, summer rainfall, maximum temperatures, snow cover and wind.

The hardiness areas have been divided into ten zones. The one marked 0 is the coldest. Other zones are progressively milder, to 9, which is the mildest. Users should locate their own area on the map and so establish the zone in which the plants are to be grown. Sometimes, even though older plants are hardy, young plants of many species may be tender and need protection the first winter.

Small areas with peculiar microclimates often exist within a zone. These areas are colder or milder than the surrounding area. They are usually too small to locate on the hardiness map or they may not have been recorded. In addition, sharp changes in elevation, as found in mountainous or hilly regions, cause a difference in climate that cannot be accurately indicated on the map. The user should also remember that the zone lines are arbitrarily drawn and that the zones merge gradually into each other. Consequently, conditions near the border of one zone may closely approximate those of an adjacent zone.

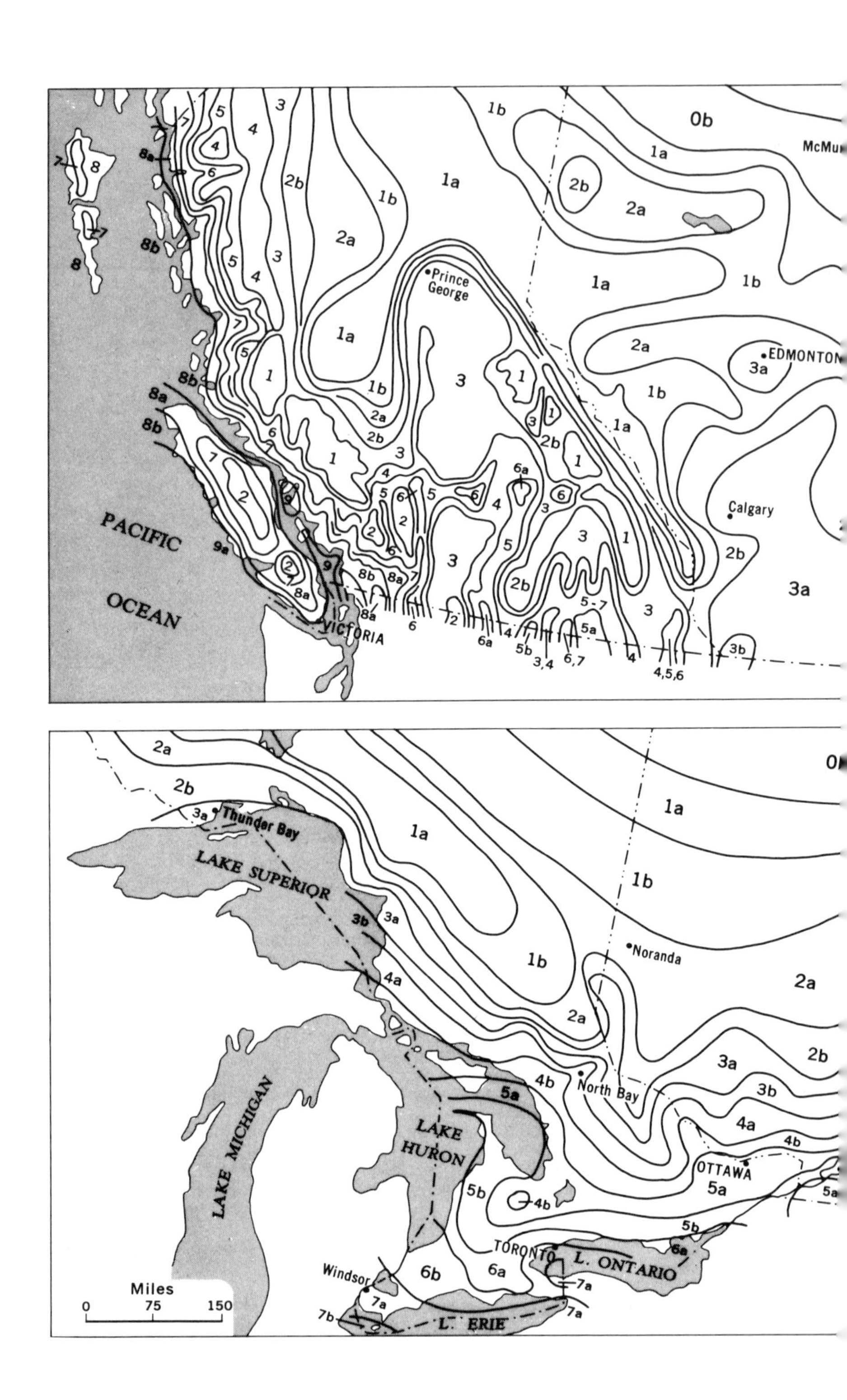
PACIFIC
OCEAN
Prince George
EDMONTON
Calgary
VICTORIA
McMu
Thunder Bay
LAKE SUPERIOR
LAKE MICHIGAN
LAKE HURON
Noranda
North Bay
OTTAWA
TORONTO
L. ONTARIO
Windsor
L. ERIE
Miles
0
75
150

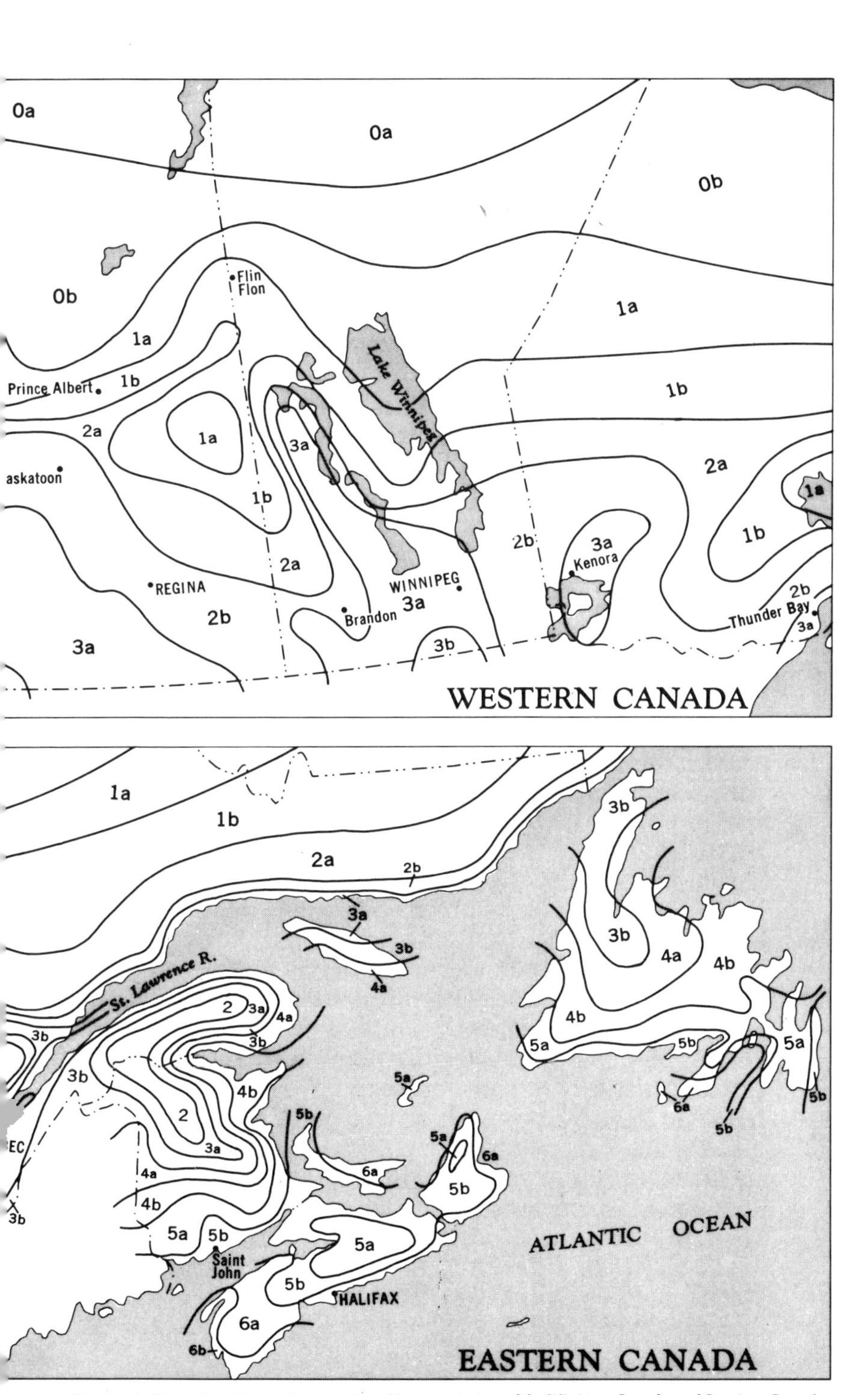

This Agriculture Canada map is reproduced by permission of the Minister, Supply and Services Canada.

Window Boxes

If your apartment doesn't have a balcony or patio, perhaps you can fix a window box securely to at least one window ledge. A good size for a window box is 1 meter (3 feet) in length, by 30 cm (12 inches) deep and 30 cm (12 inches) wide. Use the same soil mix discussed earlier in the chapter for this type of container.

Again, drips may be a problem, as the person below can complain to the manager and put an end to your garden. To prevent this from happening, line the window box with plastic and fill with a mixture of gravel and peat into which flower pots are plunged. Smaller pots, preferably clay ones, can be used, as the peat will help retain moisture. Another advantage of this method is that if a plant should finish blooming or die, the whole pot can be lifted out and replaced without causing any root damage to the other plants.

Color coordinating can be fun with window boxes if you choose flowers which contrast strikingly with the wall behind them. If your apartment block is gray, try pink pelargoniums, pink petunias, dusty miller and a little lobelia. With a white background, you might plant red pelargoniums, blue lobelia or red impatiens. Yellow marguerites, tagetes, marigolds and *Lantana camara* provide color contrast with walls of natural brick.

Hanging Baskets

Another type of planter that does well is the hanging basket. But if this is your first effort at container gardening, be warned: they are a little difficult. The main reason is that they require so much attention, mostly in the form of daily watering. The size of the basket is also important. As with other containers, the absolute minimum size is 30 cm by 30 cm (12 inches by 12 inches). Up to a point, larger is better, as the greater the amount of soil there is in the basket, the longer it will retain moisture.

Wire baskets are fine lined with moss. I always place a saucer inside the bottom of the basket, after mossing and before filling with soil; it acts as a water reservoir. The same basic mix used for planters is fine for baskets, with perhaps a little more peat added for moisture retention. There are also several products available on the market which look like little crystals. When added to soil, they take on many times their weight in water. I have friends who have used them with great success. Also, if you plan on having several baskets, I would recommend looking into one of those modern automatic watering systems that can be fitted to a regular hose outlet.

When planting up a hanging basket in early May, try to use individually potted plants in 10-cm (4-inch) pots, so that they are already a fair size. Once tapped out of their pots, plant them thickly with their roots touching each other. This way you will get a really showy basket.

Here are a few important tips about hanging baskets. Remember that drips from a freshly watered basket can be annoying to the people below. If possible, hang your baskets above other planters, which will have drainage trays. That way the drips will help water the pots below! Ensure that your baskets are securely fastened, because strong winds can be hazardous. Keep deadheads picked off on a regular basis. Last but not least, feed the plants in hanging baskets more often than plants in other balcony containers. I would suggest once every three days.

Some people purchase wooden hanging baskets which are already mature and blooming in April and May. It is most important that the plants be watered frequently and fed about every third day, as they are so root-bound by the time June arrives that they quickly deteriorate from dryness and lack of food.

Next to water, the most important item is feeding. Once a hanging basket is full of roots and well established, water at least once a day and feed with a weak solution of your favorite liquid fertilizer, whether it is fish or seaweed fertilizer, or 20-20-20. This frequent feeding is important because every time you water, you leach out the food in the soil.

Good hanging basket plants are trailing fuchsias, such as 'Lolita', 'Swingtime', 'Display', 'Flying Saucer', 'Bridal Pink' and 'Fluorescent'. It is quite easy to tell whether these plants trail or not when you look around the garden shop. Choose plants that are already growing over the edge of the pot. Nepeta (trailing catmint) is also a great plant for hanging baskets, while trailing lobelia is a traditional standby. Trailing geraniums are good, as are multiflora or trailing petunias. *Campanula isophylla* is also great in a basket; many of you may know it as star of Bethlehem. Many herbs can be grown in baskets; the thymes and parsley work especially well. Any plant that will trail or tumble is a good candidate. Of course, you can experiment with many trailing plants.

Miniature Gardens

Another area of balcony and window sill gardening that many people enjoy is the miniature garden, in which alpine and miniature plants are grown. In Europe, traditionally these gardens were made in old stone troughs. In my childhood, when an old stoneware sink was removed from the house, instead of being discarded, it was transformed into a garden. Today there are many attractive shallow planters on the market. But remember the problems with drying out and do not use anything shallower than 30 cm (12 inches).

The area of the country in which you live will have a direct bearing on the type of plants you choose. Visit your local nursery or garden shop to see if miniature plants are available. Because the varieties used are shallow-rooted, you do not need more than 15 cm (6 inches) in soil depth, but good drainage is essential. Drainage holes should be about 20 cm (8 inches) apart on the bottom, topped with a layer of rocks or crumpled styrofoam cups to stop the soil from going through the holes. A good soil mix for miniatures and alpines consists of one part sterilized soil, one part fine sand, one part moist leafmold or peat, and one part granite chippings or pea gravel. Mix everything well and pick out any large lumps or rocks. When you fill the container, make sure the soil level is a few centimeters (1/2 to 1 inch) below the rim, for watering.

The whole idea is to try to create a miniature landscape with strategically placed rocks and little evergreens. The plants listed here are of varying hardiness. Remember that although alpines come from high cold areas, they are not used to winter exposure and damp, because they generally have a blanket of snow several feet thick for protection. In colder areas, winter protection—such as chicken wire and leaves—will be required.

Mini Plants for Mini Gardens

Bulbs

Crocus chrysanthus is an early crocus about half the size of the ones most people know. Two good cultivars are 'Moonlight', which is sulphur yellow, and 'Blue Pearl', a pale blue.

Iris reticulata is a delightful miniature replica of Dutch iris. This variety produces flowers that are deeper blue on 20-cm (8-inch) stems.

Galanthus nivalis, commonly known as snowdrops, are those little white, almost bell-shaped flowers hanging on 10-cm (4-inch) stems. They are one of the first flowers of spring.

They look good at the base of a dark green miniature evergreen shrub.

Plants

Arabis alpina (rock cress) has silvery-gray foliage and white flowers. It roots easily from cuttings.

Armeria maritima (thrift) is a small light clump of grass-like leaves with 10-cm (4-inch) stems bearing bright pink clover-like flowers.

Aubretia deltoidea (purple rock cress) forms dense clusters of purple to magenta flowers early in the year. When it is finished flowering, cut it right back with scissors.

Campanula carpatica (bellflower) produces delightful bluebell-shaped flowers on 25-cm (10-inch) stems that are quite eye-catching when in full bloom. It is propagated by seed and division.

Dianthus alpinus is a minute relative of the carnation, with single rose-pink flowers. It can be propagated by division.

Helianthemum nummulari (rock rose, sun rose) forms a small bush-like plant that blooms in white, yellow, orange and red shades. As the common name suggests, the flowers open wide on sunny days.

Penstemon scouleri is another bushy, almost woody, plant which forms 15- to 20-cm (6- to 8-inch) bushes with deep lavender tubular flowers.

Phlox drummondii (rock flame) forms dense clumps of very fine foliage, covered in spring with wide open, pale pink flowers that are star-shaped.

Sedum acre (stonecrop) has small waxy leaves that are almost succulent. The flowering stems are usually about 10 cm (4 inches) long, with yellow star-shaped blooms.

There are so many more alpines and miniature plants to choose from; I would advise you to contact your city botanical garden or local garden centers for the ones available in your area. If you are an incurable gardener and have room on your patio, you can always sow alpines from seed.

There are also miniature trees available to complete your little landscape. And let's not forget miniature roses, although in areas where winters are severe they may have to be replaced on an annual basis. Nevertheless, I have seen some very attractive miniature rose gardens.

I have also seen some very successful alpine trough gardens in the Maritimes and in central Canada. I think it is fair to say that they are a challenge in areas where summer temperatures reach 30°C (86°F) and above, even though alpines enjoy full sun. In that kind of heat, I believe some temporary shading would be a good idea.

The winter temperatures are a little easier to deal with. In areas where below-freezing temperatures continue for days or months at a time, build a large chicken-wire frame around your miniature garden. Leave a space of about 30 cm (1 foot) around the edges and at the top, so that you can fill it with dry leaves as you gather them in the fall. You may have to gather leaves from the boulevard on a daily basis for a few days, and your neighbors might talk about you. But those alpines will be snug all winter and love you for the blanket. I have even seen tiny miniature gardens overwintered in large styrofoam picnic coolers.

Pests and Diseases of Outdoor Plants

Pests and diseases occur in the most carefully tended gardens. Spraying can be a problem, as the chemicals drift to the next balcony or patio. I am very aware of organic growing methods and many of the ideas are excellent. But I am sorry to say, from my own experience, that planting onions and marigolds between other plants on my balcony didn't keep the bugs away. In fact, I got spider mites and leaf miner on my marigolds.

However, when growing edible and ornamental plants

in an area as small as a patio or balcony, it is still quite easy to adopt totally organic methods of control.

One of the joys of such a small area is that it is usually the first place you go in the morning to check on what's budding or blooming. At this time, if you notice little clusters of eggs on the underside of a leaf, or a caterpillar munching away, or aphids—the finger and thumb method of destruction works like a charm. If you are a squeamish person and can't face up to it, then wear rubber gloves or use a tissue.

The following pests seem to be the most common in balcony gardens.

APHIDS can be green or black, and sometimes other shades in-between. They really can be beaten by constant, every-other-day spraying with good old-fashioned soapy water—and I mean soap, not detergent. Aphids breathe through holes in the sides of their bodies and the soap will effectively block these holes and suffocate them. However, eggs hatch out despite spraying, so the treatment must be repeated regularly.

CATERPILLARS will attack any juicy leaf on which their parents happen to have laid their eggs. The damage is easy to spot, but the caterpillar isn't. It is green like the leaf, and is sometimes hard to find. However, you will have to spot it and pick it off. Caterpillars are hard to kill with any pesticide and there is absolutely no point in spraying the whole plant.

LEAF MINERS make plants look very ugly if they are allowed to establish themselves. Marguerites and marigolds seem to be most badly affected. A little insect lays its eggs on the surface of the leaf and as they hatch, the little worms go inside the leaves and tunnel around. As soon as this is seen, pick off the affected leaves and destroy them. Keep this up and eventually the problem will go away.

POWDERY MILDEW is not an insect problem, but a disease. It is quite bothersome in high-humidity areas, especially when you try to grow tuberous begonias. One of the effective ways to control it is to have susceptible plants in areas where the air circulates freely. When the little powdery white spots appear on the leaves, spot-treat them with a mixture of powdered charcoal and sulphur dust.

SLUGS can be a problem for townhouse or condominium container gardens. You may be able to control them by handpicking them at night, but an easier solution might be to make a slug trap, using recycled plastic margarine or yogurt containers with lids. Cut three or four little holes on the side of the container near the base, so that when it sits in your container, the entrances are at slug level. Then put some bran-type slug bait in the bottom of the container, in a little pile in the center, and place the lid on. The lid prevents pets from eating the bait and keeps the bait dry. (However, you should not use this method around small children who can pry the lid off and eat the bait. It is very poisonous.) Once the container is full of dead slugs, it can be thrown out with the garbage.

SPIDER MITES will attack anything and thrive in dry conditions. First of all, I should clarify that they are not spiders, but tiny mites invisible to the naked eye. After they establish themselves near the growing tip of the plant, they start to move around on tiny webs. Daily spraying of the foliage with water from a spritzer bottle first thing in the morning will prevent spider mites.

WHITEFLIES are like little white-bodied, white-winged aphids. They love pole beans and tomatoes. Once established on plants they are very mobile, and they are not easy to control. Quite honestly, if the infestation is very bad, I would pull up and destroy the affected plants. Don't let it get to that stage, however. Destroy eggs as they appear. Another method is to go out at night and collect some of the affected leaves and pop them in a blender, quickly add water and whisk them around. Strain the juice, dilute with water in your spritzer and spray the affected plants. I know the idea of putting bugs in a blender sounds awful, but it really does work for all aphids. So don't throw out the old blender—or don't let your mate catch you using the blender, and all will be well.

II

Indoor Plants

Having dealt with ornamentals on the outside, let's move into the apartment or townhouse and talk about indoor plants. Whereas north-facing balconies are not the best settings for outdoor gardens, north- and east-facing windows are ideal for the majority of foliage houseplants. In south-facing windows, direct sun will burn even the most sun-loving plants, so during summer draw the drapes between your plants and the window during the day.

Very often rented apartments have white walls and white drapes, which make an ideal background for all plants. Many of us go in for unit furniture to house books, a TV and assorted miscellaneous items. These units can also be very attractive filled with plants! To do this successfully, artificial lights are needed. Suitable plants are prayer plant, peacock plant, snake plant, African violets or, in fact, any low-growing houseplants that can't tolerate full sun. Check the following suggestions.

Bathroom and Kitchen Plants

Ferns, anthurium, *Begonia rex*, prayer plant, kangaroo ivy, African violet.

Plants for Dark Corners (with some artificial light)

Cissus antarctica, *Ficus pumila* (climbing fig), *Calathea makoyana*, smaller species of philodendron, *Dracaena* species.

Office Desk Plants

Prayer plant, *Calathea makoyana*, *Peperomia* species, *Chlorophytum* species, *Fittonia comosum*, *Pilea* species, *Crassula argentea* (jade plant).

Plants that Grow Anywhere

Sansevieria species, cacti, *Bryophyllum tubiflorum*, aspidistra.

Hanging Planters

Often you cannot screw hooks into the ceiling, but a ledge across a window is ideal. If the view isn't great, dot it with plants. The following varieties will grow well in indoor hanging planters: *Asparagus sprengeri*, *Chlorophytum* species, *Philodendron* species, *Maranta* species, ivy.

Potted plants are used extensively these days in offices and public spaces in downtown areas. It is interesting to note that the houseplant boom we are now experiencing happened before in Victorian and Edwardian times. In those eras, only the rich could afford plants and they built elaborate conservatories in their homes with tile floors on which water could be splashed liberally to keep the humidity up. Today, if we could raise the humidity so that the furniture started to rot, then the plants would be happy!

The number-one problem encountered when growing plants indoors, particularly in the winter, is the lack of humidity. This can be eased by constant misting but that is not enough. The best way to overcome the problem is to get something like a large plastic drainage saucer or an old meat tray that is still waterproof, put in 3 to 5 cm (1 to 2 inches) of pea gravel with a little charcoal mixed through it, then keep it constantly filled with water to just below the surface of the gravel. Stand your pots on top. They will still need to be watered in the normal way, but the moisture below the gravel evaporates all day and night and will improve ailing plants by 50 percent.

From the standpoint of humidity, the bathroom or the kitchen have to be the best areas for growing plants. But very often in apartments those areas do not have a window, so artificial light will be necessary. A shelf of plants underneath the wall cupboards above a countertop can really liven up the area. In the bathroom—a good place for ferns—a whole series of shelves and lights can be set up. I must stress that ferns are highly susceptible to chemical spray damage and if the bathroom is a small area (like most apartment bathrooms), hair-spray, deodorant and other aerosols can cause ferns to lose their fronds.

I am often asked how many times a week houseplants should be watered. Because we are going through such a systemized period in history, the easy answer is: every second day. However, this is not always correct. If the soil looks dry on top, always stick your finger about 2 to 3 cm (1 inch) into the soil. If you feel moisture there, then it probably doesn't need watering. It is important to feel the soil mix before watering, as plants also wilt as a result of overwatering. In that case, adding more water to the swamp will mean sure death. It is possible to bring a dry plant back to health, but it's not as easy to revive one that has been overwatered. This is because most of the feeder roots will have died.

A great tip, which I learned from friends who grow lots of houseplants, is to keep a piggyback plant as an indicator plant. It wilts quickly when dry and is very obvious. (It also comes back fast.) When the piggyback plant wilts, check all your other plants for dryness.

Of course, there are exceptions to every rule. Some plants thrive on lots of water, such as gardenia, hibiscus, miniature citrus, fuchsia and some begonias. If they are allowed to become dry while forming flowers or fruit, the flowers or fruit drop off immediately. Most of these plants are purchased when they are just in bloom or fruit, fresh from an area such as Hawaii or Florida, which has very high humidity. They end up in our dry homes, which to them feel like the Sahara Desert. The first reaction from most of these plants is to lose their leaves within a couple of days. If this should happen to your plants, a clear plastic bag placed upside down over the bare stems will encourage new shoots and eventually new leaves will appear. Remove the bag gradually over a period of days to prevent further shock.

Some people use warm water on their plants; however, this is not necessary. Many English people water their plants with tea, which contains few nutrients but is not harmful. If you feel good about watering your plants with tea or warm water—that's fine. However, it will make very little difference to the plants.

Whenever any plant is really dry, give it a thorough soak in the bath, a utility sink or a bucket, even if it means submerging the whole pot in water until the bubbles stop.

One hears many stories about whether a plant should be watered from the top or the bottom. All of mine get their drinks from the top. (Perhaps that is why I am no good at growing African violets.)

Repotting

I am a product of the old school of gardening, and therefore a firm believer in clay pots for houseplant growing, because they breathe. Plastic pots are used extensively in commercial nurseries because they are lighter to handle and do not shatter when dropped. Purchased plants usually don't have drainage material in the bottom of the pot. This doesn't make for good drainage and allows the potting mix to run out of the bottom. Make sure when you are repotting to add a layer of drainage material.

Most potting mixes are about 90 percent peat these days, with some other ingredients added for cosmetic reasons. Because true soil mixes vary from region to region, it is much better to use purchased mixes for houseplants.

For easy care, always pot up houseplants in pots that have drainage holes and stand them in ornamental containers. If they are overwatered, you can then lift them out and tip away the excess water.

I have several times visited sick *Ficus benjaminas* (benjamin figs) that have been purchased for around $100. They first arrived at the apartment looking very elegant in white containers in which there was bark mulch. In all cases I was called after the young leaves started turning yellow and dropping. The plants had been watered on a two-day basis, regardless of whether they needed it, and the beautiful containers were half-full of water which was nicely hidden by the bark. It is always important to see where the water goes after the plant has had enough to drink.

Repotting should be an annual occurrence, but one should never expect a plant to last forever. Obviously when the plant reaches the 30-cm (12-inch) or 40-cm (15-inch) pot, it is no longer possible to repot. Later in the book, when I deal with propagation, I will show how to get new plants started from old ones. This helps to soften the grief some people experience when getting rid of an old plant.

Wherever possible, I think it is an excellent idea to start out by growing small plants in your apartment. This way they get somewhat adapted to the conditions in which they are expected to grow. Assuming the plants are small when you start out, never be tempted to put something like a benjamin fig that is in a 10-cm (4-inch) pot directly into a 25-cm (10-inch) pot, just because you know it is going to be a large plant eventually. It is far better to progress by 2.5 cm (1 inch) each time you repot. This is because air is important to roots and they always head for the sides of the pot where, particularly in a clay pot, plenty of air is available. If a plant is overpotted it will take five times as long to grow.

To get the plant out is a simple operation if these instructions are followed. Thoroughly soak both the plant and the pot before going to bed, or for several hours. When you're ready to repot, put the fingers of one hand across the top of the pot so that the plant is between your second and third finger. Then turn the whole pot upside down and tap the edge fairly sharply on a wooden surface. Providing you soaked it well, it should come out like a sand castle, and because your fingers are around the plant you have complete control. This is also a good time to check whether the root growth is healthy or not. It often tells you why the top of the plant is unhealthy.

If you are unsure whether or not your plant needs repotting, take a look at the roots. The emergence of roots from the bottom drainage hole of the pot does not always mean the plant is root-bound. Many roots go to the bottom because that is where most of the moisture is. However, if there is no soil visible and all you can see is solid root, then the plant is most definitely pot-bound and should have been potted on a couple of months ago. If you find that about 50 percent of the soil is showing at the bottom of the pot, then there is no need to repot. But if there is approximately 75 percent root coverage of the soil ball, then you should repot the plant. (However, if the plant was purchased only recently, this tapping out procedure could be disastrous. If the plant did not have many roots, all the soil will fall off. Wait at least six months.)

If there is a piece of drainage material deeply imbedded from a previous potting, don't worry if you cannot get it out. Never break roots up; if they die, some of the top of the plant will die as well.

The best time to repot is early spring, before new root growth starts. Always soak a clay pot at least one hour before using, otherwise it will compete with the plant for moisture. The pot, being extremely porous, always wins. Put some small rocks or broken pots, curved side up, in the bottom of the pot to aid drainage and stop soil from washing out through the hole. Then put a handful of well mixed and moistened soil in the bottom and try the plant for depth. By this I mean make sure that the soil level, when you are finished repotting, will be a few centimeters (1/2 to 1 inch) below the rim of the pot. Otherwise, it will not be possibe to water.

When the right level is achieved, fill in the gap between the roots and the side of the new pot, putting in a little at a time and firming the soil with your fingers if the gap is wide enough. If not, a pencil works well. As more soil is added, lightly tap the pot on a wooden surface, as this also helps the new soil to settle. When you reach a point 1 to 2 cm (1/2 inch) below the rim, firm the whole thing finger tight and water well. Then stand the plant in a shady area of the room and mist it over a couple of times a day. Do not water until the surface feels dry.

When I pot my houseplants, I use the same 6-3-2 mixture of soil, peat and sand that I described for the filling of patio pots. I realize it is much easier to buy ready-mixed potting mix. However, I have found that a lot of prepackaged mixes are far too peaty and therefore stay wet for long periods. If you feel the soil you buy is too peaty, add some coarse sand or perlite. This should be equal to one-third of the total volume of your original soil. Even though most soil mixes contain fertilizer, feeding is necessary after about three months because by that time the plant will have used up the original fertilizer. As a general rule, feed your plants with a weak solution of soluble fertilizer once a week during the summer and once a month during the winter.

Artificial Lighting

There is a great interest in growing plants under artificial lights. Much emphasis has been placed on grow-lights, which are manufactured by several companies and are a little more expensive than the average fluorescent tube. If you are growing plants for show or absolutely top quality, the grow-lights are a must. This is why they are used in plant shops. Grow-lights are the next best thing to natural sunlight; however, many plants can be grown successfully under a combination of "cool white" and "warm white" fluorescent tubes. Some ferns and plants that grow naturally almost in deep shade will thrive under a 60- or 100-watt bulb. However, because of the heat the bulbs give off, fluorescent tubes are best for apartment use.

When I was at school, we visited some caves in southwestern England where hydroelectric power had been installed. Where there were light bulbs, ferns were growing—up to 3 kilometers (2 miles) underground. This fascinated me, as the spores had probably been there for years. And I'm sure the same thing has happened anywhere in the world where caves have been discovered. Of course, there are no plants that will grow in complete darkness!

Whichever type of artificial light you use, make sure it is 30 to 45 cm (12 to 18 inches) above the plants. It should be on at least twelve hours each day but no more than sixteen hours. I mentioned at the beginning of this section that north-facing windows are fine for most foliage houseplants. However, lights will have to be used if they are flowering plants.

Most modern offices in downtown areas have permanent plant displays in their foyers which grow well because any air-conditioned building has its humidity carefully controlled. The plants love this. Also, the natural daylight is supplemented by plenty of artificial light. Those large displays are carefully tended by plant display contractors, so if any of the plants get sick, they are quickly replaced overnight. I have heard of many cases where owners had little luck with plants at home, but when they took them to the office, the plants flourished. (For information on growing herbs and vegetables under lights, see the section on using hydroponics in chapter four.)

Propagation, or What to Do When Plants Get Too Big

Many plants that get far too big become conversation pieces and embarassments at the same time. Examples are rubber plants that hit the ceiling with over half the bottom leaves missing, or dumb canes that reach 2 meters (6 feet) with no more than six or seven leaves on top. These plants require a large stake to keep them from falling over. I was once called to an apartment where a rubber plant had completely taken over the dining alcove. Because the owner could not bear to prune the tree, he left the dining area to the rubber plant and ate his meals in the living room!

In the case of the large-leaved rubber plant, air layering is an excellent way of propagating a new plant. It can be quite a messy business since wherever it is cut, white sap oozes out. However, if you lay down a few sheets of newspaper, that problem can be overcome.

The best time of the year to propagate is March to July. After this time, plants begin to slow down quite a bit.

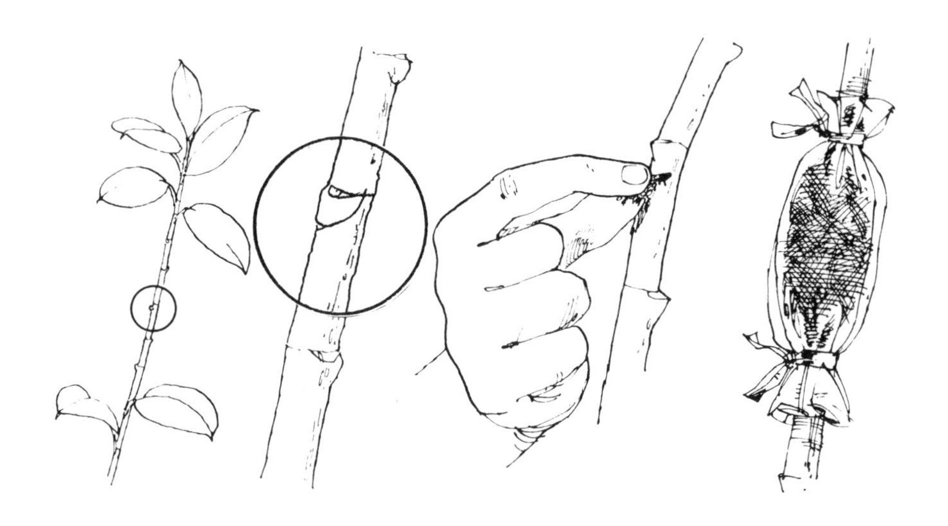

Choose a growing tip and feel back down the stem to the area where it is not yet woody but is soft enough to cut into—usually about 25 to 37 cm (10 to 15 inches) back from the tip. Remove about six leaves from this area, cutting them off fairly close to the stem. At the leaf stump nearest the center of the leafless area, or just below it, make a 45-degree cut up into the stem behind the leaf joint. Cut just half-way through the stem; do not cut the whole thing off. Then bend the stem slightly so that the cut opens up. Put a little piece of very wet moss into the opening so that it cannot heal back together again. Place a couple of handfuls of soaking moss around the whole area and wrap a clear plastic sheet about 45 cm by 20 cm (18 inches by 8 inches) tightly around it. Seal tightly at the top and bottom. The idea is to keep the moisture in.

I know it sounds as if you need three hands for that operation, but it is possible to do it on your own as long as you have everything prepared and ready within arm's length.

Keep the plant in its original home area while rooting takes place. Providing the cut was open enough, roots should appear in the moss three to four weeks later. They may take up to six weeks to fill the entire area, and when they do, cut the whole thing off just below the plastic. Remove the plastic and place the cutting in a 12-cm (5-inch) pot, or the smallest pot that will accommodate the root area. Do not try to take the moss away. Water well and mist frequently. If the plant starts to wilt, put a plastic bag over the top for a few days.

The main plant at this stage will probably look like a broom handle with a few leaves on top. If you have plucked up enough courage to air layer it, you should not feel bad about cutting it right down to .6 to 1 meter (2 or 3 feet). It will shoot out again from the base.

Another plant that gets out of hand is the *Dieffenbachia amoena*. Dumb cane is a well-tested houseplant that constantly appears in downtown offices and banks. It is usually about 1 to 1.3 meters (3 to 4 feet) high, with no more than eight leaves on top. This one is very easy to propagate, with the best results around springtime. If you thought the air layering was a pretty drastic move, the following is even more so!

All the way up the stem of a dumb cane are leaf joints (nodes) about 2.5 cm (1 inch) apart, where the old leaves dropped off. Cut the whole stem off about 7.5 cm (3 inches) above the soil level, then cut the stem into sections of two nodes, so that it ends up in little logs approximately 5 cm (2 inches) long. Dip the bottom ends of the logs in rooting hormone powder. In a thoroughly blended mixture of two parts coarse sand or perlite and one part moist peat, half bury the logs on their sides. By this I mean that one of the leaf joints should be below the surface of the rooting medium. Water well and cover with a plastic bag. Stand the whole box or pot in a warm place for rooting.

Now that we are launched into the propagation field, I would like to discuss some of the basics. In the case of everyday cuttings like pelargoniums and fuchsias, terminal shoots are taken—in other words, growing shoots. I am sure you have been told or have read that shoots should be chosen without blossoms. This is almost impossible during the late growing season, so just cut off the flowers and flower buds. A cutting should not be trying to flower when it is rooting.

Ideally, 7 to 12 cm (3 to 5 inches) is a good length for a terminal cutting. To prepare it, make a clean horizontal cut below a leaf joint, as most plants have their root initials in this area. Then remove the leaves on the bottom half of the cutting. Three to four leaves on top are enough for best results. In the case of pelargoniums, often the cuttings are prepared and left overnight so the cut ends can dry, but this is not necessary if you use hormone rooting powder.

There are several brands of hormone rooting powder on the market, often in three strengths. All are good. For the work we are discussing here, number one for softwood is ideal. None of these are magic powders that can root a broom handle, but they do help stimulate root growth and definitely reduce the risk of a cutting rotting from the bottom up. The cut is dipped in to seal it and the excess shaken off. Before these powders came on the market, we used a half-and-half mix of powdered charcoal and sulphur to prevent rotting off.

The medium in which cuttings are rooted should be very open, such as coarse sand or perlite, so that air can pass through. Many people use vermiculite on its own for rooting, but I have found that it stays too wet and turns into a solid mass when moistened. To substitute for two parts sand and one part peat, use two parts perlite to one of vermiculite or use sand and vermiculite or perlite and peat. In any event, the amount of open material should be greater than the amount of moisture retainer.

Many people root cuttings in water, but I find that a lot of the soft roots that develop in water die off quickly when potted up.

If you want to experiment with the importance of air below soil level, put a ring of cuttings around the edge of a clay pot in a rooting medium and then fill the center with more cuttings. You will find the ones at the edge root several days before the others.

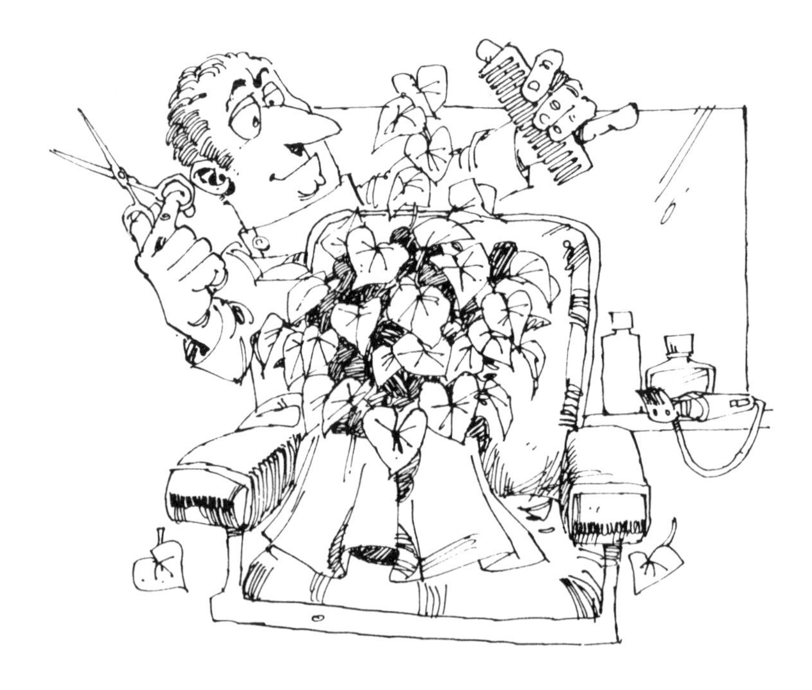

Some cuttings have enormous leaves which will be a hindrance rather than a help in the rooting process, as they take energy and sap away from the cutting. With a sharp knife or scissors, cut each leaf in half. The cut edges heal quickly. When the prepared cuttings are put into the rooting medium, make a hole for each with a pencil and bury all the bare part of the stem. That's where the roots will form—from around the leafless root joints.

Many different kinds of cuttings can be put into one pot. Water well and put the whole pot into a plastic bag. Blow up the plastic bag to keep it off the leaves, seal the top with a twist tie and stand the pot on top of the refrigerator, where there is a constant temperature from the motor. (This is similar to the temperature produced by soil-warming cables used in greenhouse propagating benches.) The top of the TV is not a good place, as it gets too hot.

Every other day, open the bag to change the air (perhaps while you are making coffee or breakfast) and feel the surface of the rooting medium to see if it needs water. Then reseal the bag. Don't attempt to move the actual cuttings for two weeks. After that time, lightly pull them; if there is resistance, you know they are rooted and can be potted up. Never pull them out and push them in again.

Not every plant has root initials just around the node. Some have them all over the stem and the leaf—for example, African violet, *Begonia rex* and gloxinia. You probably know someone who has rooted African violets from leaves. The only tip I would like to mention here is that I find they root much better in a medium other than water. If possible, let the new shoots grow to a size at which they can be handled and separated into single plants, or crowns, as these seem to flower better.

With *Begonia rex*, I was unsuccessful with the method of severing the main veins and pinning the whole leaf on a medium. The leaf area was so large and soft that it rotted off. I have had much better success with the so-called postage stamp method. All the main veins on a begonia leaf will root, so cut out the center part of the leaf with the stem on, which is cut off at 4 cm (1 1/2 inches) and used as one cutting. Cut up the remainder of the leaf into pieces about 5 cm (2 inches) square, remembering which is the base of the vein as you go.

All these squares are then planted vertically, with the base of the vein at the bottom and the leaf section one-third buried. Using the postage stamp method, twelve squares or more can be made from one leaf. Putting the leaves in vertically helps eliminate rotting, as air can circulate freely. The planting media mentioned earlier work well with *Begonia rex*. Again, put a plastic bag over the whole thing.

Many plants propagate themselves freely, like the *Chlorophytum elatum variegatum*, or spider plant. It has long stems that in turn develop plants on the ends that root readily if pinned down in potting soil. Another one is the Boston fern, which puts out all those funny little wiry runners. If they are pinned down the tips will also root.

There is often much confusion about leaf cuttings, as not everyone is sure what will root from a leaf or stem. It all comes with experience, and experimenting is great fun.

Division is another way of obtaining more plants. For example, *Asparagus plumosus* and *A. sprengeri* can be divided as long as the plant is large or pot-bound. Just tap it out of the pot, trim off most of the foliage and then cut the plant in four with a very sharp knife and repot separately. The reason the leaves are removed is that with all the severing of the roots, they would die anyway. As new roots grow, leaves will appear. Other plants that can be divided this way are *Sansevieria trifasciata* (snake plant) and aspidistra. The latter one, however, sulks for about six months before making new growth!

Having read all this, you may now feel that I treat my plants rather harshly. In greenhouses that I have tended over the last seventeen years, I have undertaken drastic pruning operations during winter months. The plants always came back to blossom freely the next summer. I am convinced that the right amount of watering, feeding, repotting and pruning is the answer. Far too many people kill plants with kindness and worry too much about the occasional brown leaf.

Miniature Dish Gardens

Dish gardens are a great way to have a mixture of plant forms and textures growing together in an attractive arrangement. They can be grown in nice old family heirlooms, like a large casserole dish. Baskets lined with plastic also look really nice. Basically any large container that can hold three to five plants or more will work.

The main difference between potted plants and dish gardens is that containers for the latter have no drainage holes. Therefore, put a 4-cm (1 1/2-inch) layer of drainage material in the bottom, and add some charcoal to keep any water that sits there sweet.

Before placing any plants in such a dish, cut a piece of garden hose the same depth as the container and place it near a convenient edge, wedging it in place with either your first plant or a handful of potting mix. Once the dish garden is all planted up, you can use that piece of hose much as you would the oil dipstick in your car. By pushing a cane down

the hose, you can determine whether or not there is excess water. It is quite simple to do and works like a charm.

The potting mix you use for repotting houseplants works fine for dish gardens. Try to choose small plants that look nice together and do not mix cactus with tropicals, as their care is quite different.

Forcing Bulbs for Seasonal Indoor Flowers

Some people like to force spring bulbs for enjoyment in an apartment from Christmas through February. This is not easy to do in all areas because of varying temperatures. To get the best selection, purchase prepared bulbs from your garden center during September/October. Hyacinths are usually a safe bet. Prepared bulbs have been specially grown over the last three years to produce a large flower and have already spent some time in a fridge to simulate winter conditions. When you plant them, they feel it is spring.

To get bulbs in bloom for Christmas, plant them the first week of October in soil mix that has been purchased, or make your own mix of two parts moist peat and one part sterilized potting soil, with a small amount of charcoal added and 15 ml (1 tablespoon) of slow-release fertilizer per 4 liters (1 gallon) of mix. Hyacinth bulbs can either be planted separately in 10-cm (4-inch) pots or in bowls with a drainage hole in the bottom.

Put rocks or broken crockery over the drainage hole and half-fill with the bulb soil mix. Then set in the bulbs so that the necks show about 1 cm (1/2 inch) above the soil surface, and leave a space for watering between the rim of the container and the soil.

Once the bulbs are planted, water well and put in a dark place that is no warmer than 5°C (42°F) for eight weeks. The really cold outside temperatures do not start until mid-November or December, so I would suggest these bulbs be put on a balcony in an old apple or pop box. If the temperature tends to get extremely low, bury the pots in peat or styrofoam. They will need to be kept moist and dark at all times without allowing frost to get to them.

The tops will start to grow in three weeks, but do not be tempted to take them out, as the eight-week dark period is necessary for root development. At the end of eight weeks, the bulbs should be brought into an atmosphere of 10°C (51°F) and good light for two weeks. This is very important for leaf growth. I am sure you have seen forced bulbs that have been brought into a temperature of 20°C (68°F). They are all leaf with the flower way down in the center. I realize that it is hard to achieve 10°C (51°F) in an apartment at the end of November, but if you want successful bulbs, you will have to work something out.

About ten days or so before Christmas, the bulbs can be brought into a room of 21-24°C (70-76°F). By this time they will be in full bloom. If, like me, you don't feel Christmas is really a bulb time, then hold them back by a later planting and have them bloom for the gray days of January, when all the Christmas decorations are taken down.

Hyacinths, tulips and narcissi are the most successful bulbs for forcing. I have yet to see a good pot of forced crocuses. If tulips and narcissus are used, bury the bulbs so that their necks do not show in the pots. Strangely enough, tulips do not send up their blooms from the top of the bulb but from the bottom and up the flat side. So always put the flat side of the bulb towards the center of the pot.

To get a really uniform display of hyacinths and double tulips, plant each bulb separately in a 7- to 10-cm (3- to 4-inch) pot, and treat each exactly the same way. When making up the bowl for bringing inside, choose plants that are the same height. Mixing colors or different bulbs all together is never very successful as the bloom is erratic.

When forced bulbs have finished flowering, cut off the deadheads, allow the plants to stay green and growing for eight weeks or so, then plant them in containers for outside. However, the flowers will be much smaller the second year.

Special Hints for Houseplants

There are many products on the market for making plant leaves shine. Very often these do more harm to the plant than good. Plants breathe through pores in their leaves and most of these shining agents, which are oil-based, quickly block the pores. If they are used too often, the leaf will drop off.

A lady once asked me why all the leaves on her rubber plant fell off. I went through all the regular questions—too hot, too dry, in a draft—all without success. Then she said that a friend had told her to put vaseline on the leaves to make them shine! Don't do it.

I use a mild nondetergent soapy water for all of my plants. Rinse them for a few seconds in a cool shower or stand them out in the rain any time when the temperature is 10°C (51°F) or above.

One problem that occurs is what to do with plants when on vacation. If the bathroom has a window, soak all your plants and stand them on several layers of wet newspaper in the bathtub. Leave the tap dripping, making sure the plug is out and the drain isn't blocked by the newspaper. If that sounds too risky, then try this solution. Water all your plants well and put plastic bags over each pot, sealing the top around the stem of the plant so that the leaves are exposed. I have a friend who puts all of her plants in a cleaner's bag, well watered and in good light. The whole thing works like a large terrarium. Although I have used all of these methods successfully for a two-week period, if I were to be away longer, I would have a friend drop by and check.

When you move into a new apartment, if possible always move the plants last. For winter moves, obviously frost protection will be required. Wrap all plants in newspaper and stand them in cardboard cartons, preferably with a top that will close over them. At any time of the year (yes, even summer), plants transported without protection will be damaged by winds of 48 to 64 kilometers per hour (30 to 40 miles per hour). Always make sure your plants are well watered a few hours before the move.

Growing indoor plants is quite a personal affair between your plants and you. Every book suggests different methods. These are some of my guidelines; the rest is up to you.

Wilting, yellow leaves; leaf drop.	Too wet.
Brown around edge of leaves and on tips; leaf drop.	Too little humidity
Wilting.	Too dry. (But be careful not to overwater.)
Leaves large, soft and dark green, beginning to curl.	Too much food
Yellowing pale green leaves; old ones drop.	Not enough food.
New shoots elongated, with large spaces between leaves.	High temperature and not enough light.
The plant looks healthy and happy. The more it grows the happier it looks.	Relax, you are doing the right thing!

Pests and Diseases of Houseplants

No matter how careful one is, pests and diseases attack houseplants fairly frequently. Great care should be taken not to touch healthy plants right after handling infected ones. Insects and diseases can easily be transferred from one plant to another.

APHIDS, usually black or green, are the most versatile little pests. They can lay overwintering eggs and if one adult finds itself on a juicy plant, it can rapidly produce live young. In good summer weather, these pests have winged generations. Even in the winter, if just one or two aphids come into the house on cut flowers, they find your plants and have a field day. They are not the most difficult pests to get rid of; often on a houseplant you can rub them off with your fingers, or they can be cleared up by spraying the plant with soapy water.

MEALYBUG is a pest that often hides right in the cracks and crevices of the stems of cacti and succulents. In other plants it appears right where the leaf joins the stem. Blobs of a fluffy white glue-like substance are a sure sign of mealybug. Because of the mealy substance on this bug's back, no spray will be effective. The only control I know is to dip the end of a Q-tip or paintbrush into rubbing alcohol and then onto the bug. Sometimes mealybugs get to such an advanced stage that it is better to get rid of the plant.

MILDEW. The only major disease in this group that affects houseplants is powdery mildew, a thin white film that can spread over the whole surface of a leaf, eventually causing the leaf to turn brown and drop off. This fungus attacks mostly begonias, but it can turn up on other plants where air circulation is poor. A fungicide or sulphur will prevent mildew from spreading, but it won't clear up the disease once it is established. Use this treatment every ten days throughout summer to prevent spores from growing on unaffected leaves.

SPIDER MITES are often prominent among the household pests. These little fellows thrive on dry conditions—in other words, the lack of humidity in an apartment. A dry window sill above a heat register is the ideal breeding ground. Just to confuse the issue, these little insects are not spiders at all, but mites, which are invisible to the naked eye.

The first sign that mites are attacking your plant is the

yellowing of young leaves as the insects suck out vital juices. Your reaction might be that the plant lacks nutrients. But even after you have fed the plant, the leaves will continue to yellow and then drop off from the growing tip. By this time, fine webs will have appeared. If you do not want to lose the plant, cut back to below the damaged area and spray with a miticide every three days over a nine-day period. On the other hand, if the plant can be easily replaced, get rid of it before it infects your other plants. The secret to preventing spider mites is to stop dry conditions by frequent mistings of water two to three times a day on healthy plants.

SCALE often appears on ferns or orchids and will spread to everything. It occurs on the underside of leaves along the main vein and is only noticed when it has a hold and the scales are brown. A close check will reveal green scales further up the vein, which house tiny mites almost invisible to the naked eye. These are doing the damage. The scales, which are quite waxy, are about .3 cm (1/8 inch) long and are the same color as the leaf when new and active. When empty, they turn brown. Scale is extremely difficult to control. Pick the scales off as you notice them, but if the plant becomes badly infected, throw it out.

WHITEFLY is a pest that has become a major greenhouse problem in Canada and therefore finds its way onto many plants headed for apartments and homes. It is particularly fond of fuchsias and hibiscus. The only effective sprays seem to be those that have a pyrethrum as a base. There are also yellow sticky boards available for placing among plants in greenhouses. The whiteflies are attracted to yellow and become stuck on the boards. If they are a major problem for you, try this in your apartment.

While on this subject, I should mention those white insects that busily wiggle around in the soil when plants have just been watered. They are the larvae of fungus gnats. They do not harm the plant at all, but live off decomposing material in the soil. Fruit flies, often attracted by moist peat, are not interested in the plant. In both cases, regular pesticides can be used to kill them, or just let the soil dry out now and again and they'll go away.

Sometimes mysterious things happen to a plant. For no particular reason, brown patches may appear suddenly on the leaves of a perfectly healthy corn plant. When this happens at UBC, we send samples to the Plant Pathology Department. The diagnosis is usually "physiological disorder," not a very comforting or helpful answer, but that is all one can say. Often the only answer is to cut the plant right down and let it start up again.

Several times in this section I have talked about spraying. When you have only a few plants at home, it is not worth investing in a large expensive pressure sprayer. If you read instructions carefully before mixing any sprays for plants you will find that most can be applied in small amounts in your plant misting bottle. Never add more than is recommended and all should be well. Another method is to place infected plants inside a plastic garbage bag and bring it up around the plant. Then spray through the opening and seal it in the bag overnight. This is often a most effective way of controlling insects.

III

Houseplants A to Z

The following list of plants for indoor growing is a cross-section that allows for shade and light areas and warm and cool temperatures. They are listed alphabetically by their botanical names, because if you need to find out about a particular plant, the correct name must be known. Common names are fine, but they vary widely from region to region.

Adiantum cuneatum
(maidenhair fern)

Adiantum cuneatum (maidenhair fern) This fern was first discovered in 1820 in Brazil, which immediately suggests a warm and humid climate. Most ferns can tolerate fairly low light intensity, which makes them ideal subjects for growing under ordinary lights, such as the ones in an apartment bathroom. However, the light must be placed about 30 to 45 cm (12 to 18 inches) from the plant, and a minimum of twelve hours of artificial light is required per day. Maidenhair ferns like a soil that resembles a woodland type with two-thirds peat or leafmold. Good drainage in the form of coarse sand or perlite is important. Very often the fronds tend to turn brown during winter months. If this happens, cut the whole thing back and let it start afresh. Scale is one of the major pests of ferns. If your plant becomes badly infected, throw it out before the scale has a chance to infect everything. There is no easy control other than picking off scale as it occurs.

Agave americana
(century plant)

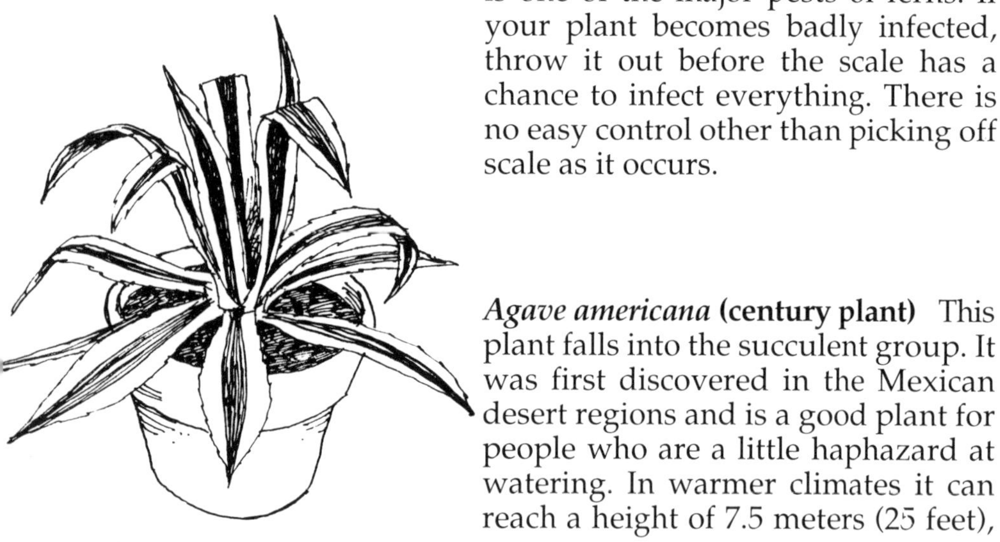

Agave americana (century plant) This plant falls into the succulent group. It was first discovered in the Mexican desert regions and is a good plant for people who are a little haphazard at watering. In warmer climates it can reach a height of 7.5 meters (25 feet), but the largest I have seen as a houseplant is .6 meter (2 feet). Because it has vicious spines on the ends of the leaves, this plant should be handled with care. The very handsome variegated form is seen more often in cultivation than the green type. Century plants are tolerant of poor soil conditions, so the best potting soil need not be used. Water once a week if possible. Always use a clay pot for all types of succulents; sitting in wet soil is not their strongest point.

When these plants become potbound, the older leaves tend to turn yellow, then black; pot up to the next size when this starts to happen. Hot dry apartments are ideal places for century plants. Mealybug and scale are the two major pests.

Just a word about the myth that surrounds the century plant: it isn't true that it has to be 100 years old to bloom. Some species bloom after six years, while others take thirty to forty years.

If you have a large sunny balcony and live in an area where the summer gets fairly warm, this plant will survive well outside during that season. A century plant will add a Spanish flavor to your patio.

Aglaonema commutatum
(Chinese evergreen)

***Aglaonema commutatum* (Chinese evergreen)** This plant was discovered in the Philippines in 1863. It looks very much like a rather small and plain green dumb cane (*Dieffenbachia amoena*). I have found that it is one of those plants that is hard to kill and it tolerates quite wet conditions. Like all plants in this group, Chinese evergreens look much better when they are planted several to a pot. Poor light conditions don't seem to worry it too much.

This plant does best in regular well-drained potting soil, but I have seen it growing in pebbles and water with fertilizer added at regular intervals. Spider mite, mealybug and scale are the main problems with Chinese evergreens.

Araucaria excelsa
(Norfolk Island pine)

***Araucaria excelsa* (Norfolk Island pine)** This beautiful houseplant has airy light green branches that ray out horizontally from a central stem. It got its common name from the island where it was first discovered. There it attains a height of up to 60 meters (200 feet), but don't be alarmed. The largest specimen I've seen in cultivation was 2 meters (6 feet) high.

Good humidity is the key to success with this plant. High temperatures are not required—18 to 21°C (65 to 70°F) is the ideal range. Care should be taken not to let a Norfolk Island pine stand in the draft or near a heat register. This can quickly turn a whole layer of branches brown.

A fairly rich potting soil is required with adequate drainage, and a regular feeding program through the year is necessary. It is much better to start off with a smaller plant and pot on, because once a height of 1.5 to 2 meters (5 to 6 feet) is reached and the plant begins to starve, it is difficult to prevent the lower branches from dropping off. Spider mite is the main pest of Norfolk Island pines, so mist regularly.

Asparagus plumosus (wedding fern)

***Asparagus plumosus* (wedding fern)** and ***Asparagus sprengeri* (asparagus fern)** These plants are both referred to as ferns because of their light and airy foliage, but they are not true ferns. True ferns multiply from spores; these plants are in the lily family and produce minute flowers followed by seed. (However, this doesn't happen very often in an apartment or house.)

Asparagus plumosus was discovered in South Africa in 1898 and it gains its common name because it is always used in bouquets. *Asparagus sprengeri* is from Natal. Its fronds have a more cylindrical form and the plant is quite prickly.

Aspidistra lurida
(cannon ball)

Both are excellent substitutes for true ferns, as they can survive well in dry conditions and do extremely well in hanging pots. Being of the lily family, they have very strong tuberous root systems and quickly become potbound, causing whole fronds to turn yellow and eventually die. When this starts to happen, it is better to cut off all the top growth, tap the plant out of its pot and divide it by cutting the whole plant in half from top to bottom. Then repot the two halves singly. These ferns are happy with just about any soil. Their main pest problem is spider mite.

***Aspidistra lurida* (cannon ball)** This was a popular plant of the Victorian and Edwardian eras. Because it was such a hard plant to kill, it survived for generations and was passed on as an heirloom. Gracie Fields used to sing of "The Biggest Aspidistra in the World." The plant was so common that people became sick of it. However, with the recent plant boom it is regaining popularity and rightly so, because it can stand drafts, damp and dusty conditions. It originally comes from China, where it was discovered in 1822. It blooms each year, often without being noticed, as the strange, waxy, pinkish white flowers occur at soil level.

As already suggested, this is an easy plant to care for, since it is not fussy about soil or light conditions. An aspidistra is easy to divide, but it can take six months to get over the shock and finally start to send up new leaves. Once I saw an aspidistra with spider mites, but as a general rule it is a pest-free plant.

Begonia masoniana
(iron cross begonia)

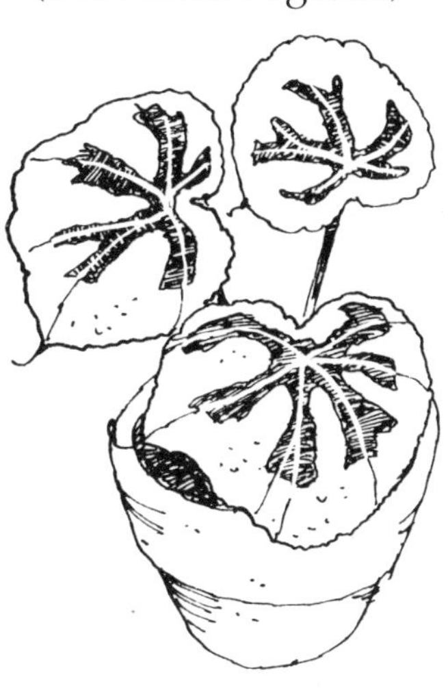

***Begonia masoniana* (iron cross begonia)** Introduced fairly recently, this plant is now famous, appearing in many glossy houseplant pictures. It is named for a Mr. L. Maurice Mason of Singapore Botanical Gardens, and it was probably discovered in Indochina. The leaf form and culture is very much like that of the *Begonia rex* mentioned below, but it has a much more uneven surface of emerald green with distinct dark markings near the center that suggest an iron cross.

Begonia rex
(begonia)

***Begonia rex* (begonia)** This is one of the begonias grown for leaf form and color. There is quite a large collection in this group, ranging from silver to pink to red foliage. The flowers on both the begonias mentioned here are terribly insignificant. Rising from the rhizomatus stem, they are most often

Bryophyllum tubiflorum
(air plant)

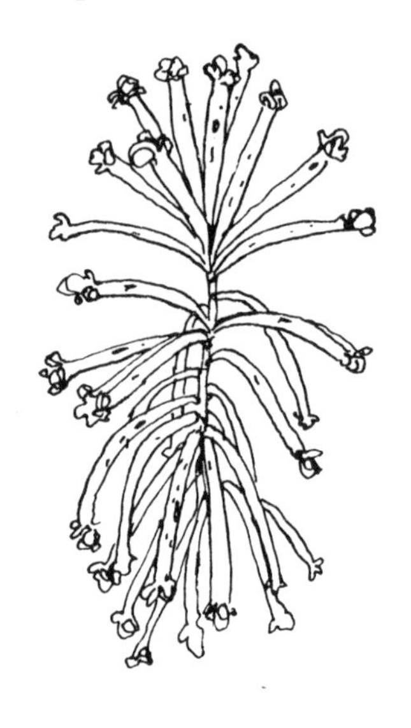

completely covered by the leaves.

Begonias make tremendous bathroom and kitchen plants because they love high humidity. They are ideal apartment plants as they grow slowly and do not require frequent repotting. Their soil should be of the woodland type, about two-thirds leaf mold or peat, but with good drainage. Regular feeding is important. Strong sunlight is harmful but under artificial light the colors in the leaves become very pronounced. Propagation of both these begonias is a simple procedure, using leaf cuttings. (See the section on propagation in chapter two.)

The main problem with begonias tends to be mildew, which is a fungus disease occurring on leaf surfaces. There are many fungicides on the market, but ordinary sulphur powder lightly dusted on the leaves is a fairly good control.

***Bryophyllum tubiflorum* (air plant)**
First discovered in Madagascar, air plant has become a noxious weed in other tropical areas of the world.

This plant has fleshy leaves about 5 to 7.5 cm (2 to 3 inches) long with little plantlets all around the edges. These plantlets drop off and spread by rooting into whichever area of soil they fall. For this reason it's not one of my favorite plants; the plantlets can fall into other plant pots and spread all through them. Despite all that, the air plant can restore your faith in growing plants. It is never very bushy, but it can grow to a height of 1 to 1.5 meters (3 to 4 feet) in a 20-cm (8-inch) pot.

Because the air plant is a succulent, don't overwater, and stand it in a well-lighted area. If it gets good sun, nice orange tubular flowers will appear near the top of the plant. I have never seen a sick air plant, but I'm sure that, given the chance, mealybug would attack it.

Buxus mycrophylla
(little leaf box)

***Buxus mycrophylla* (little leaf box)**
This is a relative of the so-called English boxwood, which at one time was used for little hedges all around formal European gardens. Discovered in Japan in 1860, it has since been introduced into many areas of the world. When I first came to Canada, I was surprised to see it sold as a houseplant—few plants from temperate regions can survive the artificial warmth of a home in winter. Experience has proven that it works well as a houseplant, but it does need good light and a cool window ledge to prevent a spindly, leggy appearance.

This plant is not fussy about the type of soil it needs, as long as fertilizer is applied on a regular basis. Larger plants can be clipped into interesting shapes, and cuttings taken during June and July seem to root quite readily. It responds well as a balcony plant in summer and is an extremely good plant for cooler apartments.

Caladium
(elephant ear)

***Caladium*—various species (elephant ear)** These are most attractive plants. However, take care—these South African natives are difficult for some folks, as high temperatures and humidity are important factors. The large leaves, shaped like elephants' ears, are brightly colored or white with green veining and are carried on single slender stems about 45 cm (18 inches) long.

Never buy a plant that has been sitting outside a store in spring or fall. If you do, chances are that when you bring it indoors it will gradually collapse leaf by leaf on the dining room table because it has been chilled. Elephant ears are brought into garden shops at various times of the year from the south. When purchased, they are usually about six to twelve weeks old, and they proceed to die back gradually. As this happens, cut back on the watering over a period of three weeks or so, and after all the leaves have turned brown, put the pot on its side in the cupboard under the sink and forget about it for three to four months. Then take it out and shake the soil from the pot. You will discover a little flat corm just under the surface of the soil. Put this in a small pot of moist peat with a plastic bag over the top and place it on the refrigerator or somewhere equally warm. Check each day to make sure the bulb is not going moldy. By the time three weeks have passed, it should have rooted. As it grows, pot on from one size of pot to the next. Eventually you will end up with a 12- to 15-cm (5- to 6-inch) pot.

Regular well-drained soil is fine for elephant ears. They need the warmest, most humid area you can provide. Often the leaves show drips of moisture on their tips—a natural occurrence and nothing to worry about on healthy plants.

Calathea makoyana

Calathea makoyana This is a cultivar introduced by Jacob Makoy and Company of Belgium; however, the *Calathea* species were first discovered in Brazil and are close relatives of the prayer plant. In fact, they have the same tendency to fold their leaves at night. (See *Maranta*.)

This particular plant is definitely marked with maroon splotches, making the oval leaves look rather like peacock feathers. It remains much more

Ceropegia woodii
(rosary vine)

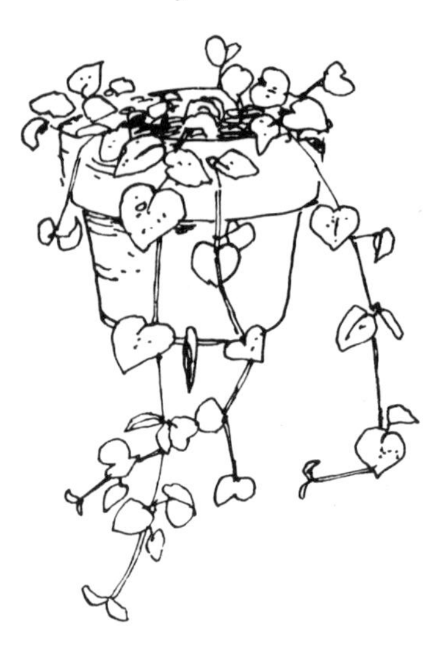

Chamaedorea elegans
(parlor palm)

controlled than the prayer plant as it does not spread sideways. I use it successfully as a terrarium plant and have one in a three-year-old bottle garden. *Calathea makoyana* is easily propagated by division and seems untroubled by pests and diseases. Odd brown spots occur if it gets too much sun.

***Ceropegia woodii* (rosary vine)** This fascinating plant is from Natal. It is a trailing plant of a succulent nature with heart-shaped leaves 1 cm (1/2 inch) long growing at 7-cm (3-inch) intervals. It makes a very good plant for hanging in a sunny window. It is extremely easy to propagate; if you place one of the runners on top of a pot of soil, it will form roots at every leaf joint.

The large lumps that occur in the soil are storage tubers. In its natural habitat this plant has to survive a long time without rain. Never allow rosary vine to sit wet for very long, as it will rot, but water once a week in summer.

***Chamaedorea elegans* (parlor palm)** This elegant plant was discovered in Mexico in 1873. It reaches approximately 1.5 meters (4 feet) in a 25-cm

(10-inch) pot. Never a bushy plant, it has a very thin stem. Strange flowers occur during summer, rather like a big spray of little yellow balls. If these mature and form seeds, it is easy to germinate enough plants for the whole apartment block.

Most palms suffer from brown leaf tips wherever they grow in the world. However, if the parlor palm is kept potted up on a regular basis and is well fed, it should remain a healthy green right to the tips. Spider mites will play havoc with this plant if they are not controlled when first spotted.

Chlorophytum elatum variegatum
(spider plant)

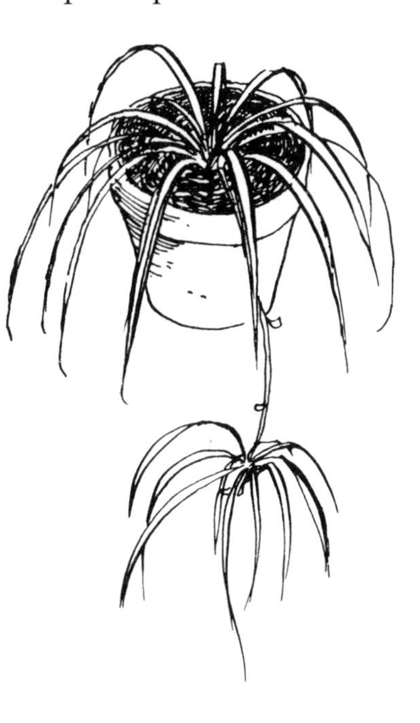

***Chlorophytum elatum variegatum* (spider plant)** This really is a must for hanging planters because of its tendency to send out long pendant runners with new plants on the end. A native of South Africa, the spider plant is in the same family as the asparagus fern and has the same root system with fleshy white tubers. I have seen one of these plants force a clay pot to crack when it became pot-bound. Keep an eye out for this and repot when necessary, sometimes as often as twice a year. The flowers of the spider plant are an attractive white with bright yellow stamens. As much as I am against using unsterilized garden soil for houseplants, this is one that I know will survive in just about any type of soil imaginable.

Some of the *Chlorophytum* species do not send out runners when the plant is being cared for too well. Sometimes the ends of the leaves turn black, but I know of no cure, as this happens even in the greenhouse where growing conditions are ideal. However, generally spider plants are trouble-free.

Cissus antarctica
(kangaroo vine)

***Cissus antarctica* (kangaroo vine)**
Quite often the common name of this plant is tagged onto *Cissus rhombifolia*, which is a different plant with different origins. *Cissus antarctica* was discovered in Australia in 1790. It is a vine with oval, pointed leaves approximately 3 to 5 cm (1 to 2 inches) long, which have toothed edges that are somewhat coarse to the touch. The plant climbs with the aid of tendrils. Kangaroo vine is an invaluable plant for poor light conditions and it always remains a shiny bright green. Because of its leathery texture it is pest-free, except for the occasional mealybug. Not only is it successful as a climber, but it also does well as a hanging plant in cool, poorly lit areas.

***Coleus blumei* (coleus)** This perennial from Java should not really be included in the houseplant list. However, on the west coast of Canada it is sold as a houseplant and everyone gets terribly upset at the end of September when the whole thing suddenly sickens and loses vigor at a rapid rate. Many of you know this plant as an annual in areas of the country where summer nights are warm. An attractive nettle-leaved plant with brilliantly colored leaves, it should be treated as an annual indoors as well.

It is very easy to grow coleus from seed each year, and it is also easy to grow from cuttings, especially if you have a particularly good form that you want to carry over to another year. But remember to take cuttings during August. Coddle the new plant on through the winter indoors under artificial light and discard the old plant. Aphids are a constant problem with this plant, but they are relatively simple to control.

Coleus blumei
(coleus)

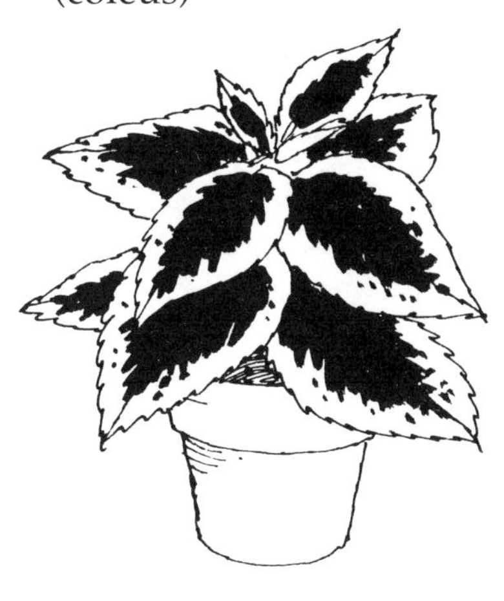

Cordyline terminalis
(Hawaiian ti plant)

***Cordyline terminalis* (Hawaiian ti plant)** Discovered in tropical Asia in 1749, this is not the easiest houseplant to grow, but it is so attractive that it is worth the work. Its common name is somewhat confusing as there are several species sold under the same name.

This particular plant has a 15- to 30-cm (6- to 12-inch) main stem topped with a cluster of leaves, 20 cm (8 inches) long by 5 cm (2 inches) wide, with the leaves coming to a point, in beautiful shades of maroon and pink. Like so many plants that lack chlorophyll in their leaves, this one is very prone to leaf deterioration, mostly in the form of large brown areas. The browning seems to happen most frequently in summer, if too much greatly magnified sun is allowed to shine on the plant.

Because the ti plant is a terminal-shooted plant it's a good idea to put three plants together in one pot for a more pleasing effect. The stem roots easily, using the same method as for dumb cane. (See the section on propagation in chapter two.) Regular feeding is very important.

Crassula argentea
(jade plant)

***Crassula argentea* (jade plant)** This popular North American houseplant is a succulent from South Africa. It is relatively easy to grow, but people do run into problems—the main ones being overpotting and overwatering, which makes the root rot.

I know of one specimen that is twenty-five years old and quite large. However, it has to be propped up between two sofas, as it is top-heavy.

To keep the growth as compact and sturdy as possible, place the plant in an airy part of the apartment with a temperature of about 18 to 21°C (65 to 70°F) and water only when dry. I feed mine only once a month and stand it outside on sunny summer days. As far as propagation goes, it can be rooted easily from regular terminal cuttings. Also, each leaf will root. Mealybug can be a major pest of the jade plant, so keep a close watch on it.

Cryptanthus bivittatus
(earth star)

Cryptanthus bivittatus **(earth star)**
This plant is well named, because it has a star-like appearance. In fact, grouped in a terrarium or dish garden, they can look rather like a cluster of starfish.

The leaves are sharply pointed, arching and spiny along the edges. They are greenish brown in color, and when in humid conditions, such as a terrarium, they have beautiful reddish pink stripes down the middle. They also bear star-shaped white flowers in the center of the plant. Earth star belongs to the large family Bromeliaceae, which has many interesting species worth investigating for indoor gardening.

Cyperus alternifolius
(umbrella grass plant)

Dieffenbachia amoena
(dumb cane)

***Cyperus alternifolius* (umbrella grass plant)** This plant is a must for anyone who is heavy-handed on the watering. A swamp plant, it was discovered in Madagascar in 1893.

As umbrella grass plants grow fairly rapidly, they need to be divided often. Propagation often occurs naturally by seed. A fascinating way to propagate is to bend one of the stems over so that the center of the umbrella part at the top touches water—such as a tumbler full of water. Roots will appear in two to three weeks. I have seen these plants grown successfully in pebbles and water with fertilizer added. Even though it is a moisture-loving plant that should house enough water to keep spider mites away, that pest is often a problem. If you have an umbrella grass plant that you water only when dry, that's fine. Don't suddenly start to add more water unless the plant looks sick.

***Dieffenbachia amoena* (dumb cane)** This plant gets its common name for a very good reason—it is a poisonous plant if eaten. This does not mean it should be discarded. It just means that you should be careful if you have children or pets around that are liable to chew leaves. The poisonous effect is that one cannot talk for a week or so, as there are some calcium oxalate crystals in the leaves that cause the throat to swell. The swelling can be very painful, depending on the amount ingested. I very often get calls from people whose cats have eaten some dumb cane.

This plant is attractive if grouped, but it can get unsightly on its own when it is 2 meters (6 feet) high and has only seven leaves, with a great stake holding it up. At this stage it is time to propagate. (See the section on propagation in chapter two for instructions on how to propagate from the stem.)

Scale and mealybug can be a problem with dumb cane. As far as soil goes, any good potting soil is fine, providing it is sterilized.

Dizygotheca elegantissima
(false aralia)

***Dizygotheca elegantissima* (false aralia)** First collected from the New Hebrides in 1873, in recent years it has become very much one of the trendy plants for interior decoration. The palm-shaped leaves of this plant are a beautiful darkish red green and are deeply toothed around the edges. I would suggest you purchase small plants and pot two or three together for a fuller display, as false aralias never bush out.

If a large plant is purchased it can

Dracaena surculosa
(gold spots)

be very susceptible to leaf drop overnight, resulting in a completely bare stem. If this happens, do not despair. Mist the stem and put a clear plastic bag over the whole thing until it starts to sprout again.

These are not easy plants to propagate except in tropical regions where they are grown from seed. Spider mite is its number-one pest.

***Dracaena surculosa* (gold spots)** Many *Dracaena* species are dealt with at length in other books, but this variety is quite different. It has leaves in pairs on bamboo-like stems and it is quite bushy. This variety was discovered in tropical Africa in 1893. I once saw a gold spots produce flowers in an apartment. They were in clusters, greenish white and sweetly scented. The occasional plant sets bright orange berries.

Like all other *Dracaena* species, this type is tolerant of many different potting mixes and enjoys being potbound. Poor light doesn't bother it either. Because of the leathery leaves, cuttings root extremely easily and few bugs attack the plant.

Euphorbia pulcherrima
(poinsettia)

***Euphorbia pulcherrima* (poinsettia)** I have to include this in the list as so many people wish to know how to make it red for Christmas and how to keep it from year to year.

Poinsettias are native to Mexico, where daylength and nightlength are almost equal. Poinsettias must have twelve hours of complete unbroken darkness each night from the first week in October to the first week of December. In an apartment this usually means a cardboard carton with a dark cloth over the top. But please remember to take the plant out for twelve hours of good light each day.

In February, when it finishes flowering, cut the whole plant back to 15 cm (6 inches) and continue to water and feed. As soon as it is warm enough at night (around June), put it out on the balcony for the summer. Cut back again in August and root the tips if you wish. Then bring it in the last week of August or the first week of September, depending on the weather. Pruning time is quite critical as pruning later than August will dispense with flowers. Regular feeding right through to Christmas is essential.

Fatsia japonica
(fatsia)

***Fatsia japonica* (fatsia)** This handsome large-leaved plant was first introduced from Japan in 1838. The leaves resemble large ivy leaves and are dark green and leathery. Healthy plants

Ficus benjamina
(benjamin fig)

produce rather pleasant whitish green globular flowers in clusters during the summer. It is one of those versatile houseplants that seems to be happy outside on the patio for the summer and inside the house during the winter.

When restricted to a small pot, fatsia becomes quite starved, so a year-round feeding program should be followed. A rich potting mix and regular repotting are also important. Large plants may be pruned back without causing any harm to the plant. Scale can be troublesome; if the plant becomes badly infected, throw it out and start again.

***Ficus benjamina* (benjamin fig)** Originally from tropical Asia, this is one of those plants that can cause a lot of heartache. The larger a plant is, the more it tends to lose its leaves when brought into the house or apartment. Therefore, good light and humidity is important. Never allow it to get too wet, as this also can cause rapid leaf fall and sudden death. It is far better to give the plant a good drink when dry and then allow it to dry out again. If it is in a pot that is standing inside another container, be sure to check the water buildup at the base of the pot. If a large plant does lose leaves, mist it regularly and try putting a large plastic bag over the whole thing.

This particular *Ficus* has little waxy deposits on the leaves, which often cause alarm. They are produced by the sap of the plant oozing through the pores in the leaves—a natural function which does no harm to the plant.

Ficus pumila
(creeping fig)

***Ficus pumila* (creeping fig)** This is a delightful member of the fig family which lends itself well to all facets of indoor gardening. Its oval, bright green leaves, about the size of a dime, grow along attractive wiry stems. It climbs by means of tiny, hair-like roots on the underside of the stems, much like ivy.

This plant is ideal for hanging in a cooler north- or east-facing window. You can let it climb the walls of your apartment, but be warned: it marks the walls. Keep it well misted to prevent the buildup of spider mite.

Fittonia Verschaffeltii
(nerve plant)

***Fittonia argyroneura* and *Fittonia Verschaffeltii* (nerve plant)** *Fittonia argyroneura* is a delightful low-growing tropical plant first found in Peru in the 1800s. It is a ground-cover plant with oval leaves about 5 cm (2 inches) in length and a nice dark green laced with pink veins. The stems root easily at each leaf joint. Because of its creeping habit, it makes a nice hanging plant, except that high humidity is important. This can be a problem when the plant is hung near the ceiling, which is the hottest, driest part of a room. *Fittonia Verschaffeltii* grows much faster and has white veins. Both plants are extremely good for terrariums. They can tolerate plenty of moisture, but a well-drained soil mix is essential.

Gasteria verrucosa
(ox tongue)

***Gasteria verrucosa* (ox tongue)** This South African native has a quite peculiar form. It is a succulent with angular leaves raying from the center, all covered with little white lumps in bar patterns. The dainty white flowers are carried on long slender stems.

This plant prefers a very open or gritty, poor soil and seems to detest fertilizer. I have seen many plants rot from being too wet. Strangely enough, bright sunny window ledges are not the ideal spot for ox tongues; filtered sun is ideal.

There doesn't seem to be any particular time for dividing up and repotting. This is usually done when the plant becomes overcrowded in its pot. A good potting mix is made up of equal parts of soil, peat and coarse sand. Always use a clay pot.

***Gynura aurantiaca* (purple passion plant)** This plant has lost its popularity in recent years. Discovered in Java in 1880, it is a trailing plant with oak-like leaves approximately 5 to 7 cm (2 to 3 inches) long. All of the leaves and stems are covered with fine purple hairs, which seem to intensify in color when subjected to strong sunlight. This plant is tolerant of all kinds of soil conditions but it appreciates regular feeding. Propagation is simple, using a terminal stem cutting.

Although it is an extremely good hanging plant, purple passion plant has two disadvantages. First, it is a host plant for aphids, which means that if there are any around, they will certainly appear there first. The other problem is that the flowers have an objectionable odor, so you should pick off the unopened buds.

Gynura aurantiaca (purple passion plant)

Hedera helix (English ivy)

***Hedera helix* (English ivy)** There are many fine species of ivy in cultivation, some of which are variegated. The main thing to remember about ivy is that cooler temperatures are the key to success. Most ivies appreciate being put outside on the balcony on mild wet days and being left out all summer in semi-shaded areas. Ivy is often used in topiary work, as it is trained over shapes of chicken wire stuffed with a mixture of sphagnum moss and peat.

Young ivies from freshly rooted cuttings are very frustrating, because they just sit there doing absolutely nothing for eight months to a year, until suddenly a new growth starts with a vengeance. To grow ivy as a little tree, graft a bushy growing species onto a stem of *Fatshedera Lizei* or *Fatsia japonica*. When taking cuttings of ivy, use a long trail. Put the cut end into the rooting medium and pin the rest in a spiral on the surface of the medium. Eventually roots will form at each leaf joint.

Hypoestes sanguinolenta (polka dot plant, freckle plant)

***Hypoestes sanguinolenta* (polka dot plant, freckle plant)** It's a pity that freckle plant isn't used more widely as a houseplant—it is so attractive.

Its tiny, almost heart-shaped opposing leaves about 2.5 cm (1 inch) long are dark green and covered with little pink dots—almost as if one had splashed paint all over them. Good light is essential, otherwise the pink dots fade away. As a bonus, freckle plant bears attractive purple flowers just 4 cm (1 1/2 inches) long. In a 20-cm (8-inch) pot it will reach a height of about 60 cm (2 feet).

This plant grows rapidly and can quickly become starved. It responds well if you cut it right back, shake the soil off the roots and then repot. (I hasten to add this is not a common practice in all repotting!)

Jacobinea coccinea
(Brazilian plume)

Jacobinea coccinea **(Brazilian plume)**
This is a South American native with lovely shrimp-pink flowers and dark, shiny, red-green foliage. Brazilian plume requires a good rich soil, high humidity, and very warm temperatures. An adequate amount of filtered sunlight is also important to initiate the bloom and prevent the plant from becoming tall and spindly. After it has finished blooming, cut the plant back to approximately 10 cm (4 inches) of bare stem. Continue watering and do not allow it to dry out. This is one of those plants that you graduate to, once the initial worries of growing houseplants are over.

Jasminum polyanthum
(pink jasmine)

Jasminum polyanthum **(pink jasmine)**
This is the type of plant that needs room to climb. If you have an enclosed, greenhouse type of balcony that is frostproof, this plant is a must. In February it produces beautiful clusters of small white trumpet-shaped flowers with an almost overpowering scent.

Kalanchoe blossfeldiana

This jasmine can be fairly untidy if not kept pruned. It needs a large pot with some kind of trellis work for support. Light is an important factor, as it is with all flowering plants. If you have one that hasn't flowered, try severe spring pruning. Cut the whole thing down to bare wood. Cuttings root easily if taken through March and April.

The soil should be a rich mix with good drainage. Water only when dry and keep a close watch for spider mites.

Kalanchoe blossfeldiana A winter-flowering plant from Madagascar, in some areas this plant competes with poinsettia as a Christmas plant.

Kalanchoe grows to a height of 25 to 30 cm (10 to 12 inches). The glossy succulent leaves are in opposite pairs and are slightly toothed. During November and December, clusters of star-shaped brilliant red flowers appear. Very often if the plant is kept from year to year, it doesn't bloom well, due to poor light conditions. If you have one that doesn't flower, treat it as a poinsettia. Each day during October and November, give it twelve hours of darkness and twelve hours of warmth and light.

As far as potting soil goes, a good open mix is required, coupled with rich soil, preferably in a clay pot. Even though *Kalanchoe blossfeldiana* is a succulent, it appreciates good food.

Cuttings from both leaves and shoots work well if taken in spring. The only pest that attacks this plant is mealybug.

Lantana camara
(Jamaican verbena, lantana)

***Lantana camara* (Jamaican verbena, lantana)** Attractive and versatile, lantana is extremely useful for both a sunny balcony in summer and a cool window during winter. It was first discovered in Jamaica in 1692 and the flowers do resemble those of verbena, hence its common name.

A weedy shrub in its native surroundings, lantana is sometimes treated as a small tree in a pot. The stems are square and the nettle-like leaves are in opposite pairs. Both are covered with minute hairs which can be irritating to the touch. The stem, in particular, is quite prickly. When the leaves are bruised they emit a strong odor.

The flowers are pretty, in neat little umbels, and on the same head range in color from yellow and orange, through red to pink. In summer the whole plant is covered with these flowers. Lantana responds well to severe pruning in the winter. Prune right back to the old wood. The plant is never badly affected by insects.

Lithops
(living pebbles, living stones)

***Lithops*—various species (living pebbles, living stones)** These plants are mostly from the desert regions of South Africa and so far do not appear to be very popular as houseplants in North America. They really do resemble the pebbles that they grow amongst. They are 2.5- to 5-cm (1- to 2-inch) ovals, sometimes with a split across the top where the flowers appear. The blooms are a daisy-like white or yellow, almost as large as the plant.

Like so many of the small cactus group, living stones fare better when planted several to a pot or dish garden, rather than individually in small pots that dry out quickly and cause the plant to dehydrate. Naturally an open mix is good for these little plants. Providing they have good drainage, water them once a week, especially if they are kept on a dry, sunny window sill.

Maranta leuconeura
(prayer plant)

***Maranta leuconeura* (prayer plant)** So-named because the leaves close together at night, it was first introduced into cultivation from Brazil in 1875. It is one of the popular never-fail plants and is very easy to grow as long as the humidity is kept up. It even does rather nicely in a hanging planter, because of its low spreading habit.

After six to eight months, many

people encounter problems with what has been a healthy plant. The leaves start to turn brown around the edges; eventually they go completely yellow and do not close up at night. The plant is telling you that it needs a rest, despite the fact that it is tropical. If you have ever repotted a prayer plant, you will have noticed small lumps somewhat like miniature potatoes all over the root system. These are food storage areas and are also an indication of a plant that rests at some stage. If and when this starts to happen, reduce the watering gradually and dry the plant off for six to eight weeks, then cut off dead leaves, shake the old soil from the roots and repot. Water well and after about two weeks new shoots will appear.

Monstera deliciosa
(Swiss cheese plant)

***Monstera deliciosa* (Swiss cheese plant)** This Mexican native has enormous leaves that resemble a split-leafed philodendron, but it is superior to the philodendron because it always produces cut leaves. The only problem is that the plant becomes enormous, with long brown roots appearing out of the stem. In the jungle, where the plant grows to great heights in and along tree branches, it sends down these long aerial roots to get nutrition from the jungle floor. Indoors, the plant

Myrtus communis
(myrtle)

finds only a living room floor, and the owner administers the nutrition in the form of plant food. The aerial roots can be removed without causing any harm to the plant.

If you should have the conditions that encourage your *Monstera deliciosa* to produce fruits, do sample them—they are really delicious, tasting of pineapple and guava. (However, like many members of this family, there is an irritant surrounding each succulent portion of the fruit. If you are susceptible to allergies, don't eat it.)

The only problem I know of with this plant is scale.

***Myrtus communis* (myrtle)** Myrtle comes from a fairly wide area covering southern Europe to West Africa. The foliage is dark green and thickly clustered along the light, woody branches. Myrtle should be grown much more than it is, if only for its scent. When crushed it gives off a delightful odor. Sprigs of myrtle have traditionally been carried in royal wedding bouquets.

This plant is happy outside for the summer months but must be inside for the winter, where it is equally happy. Cuttings root easily from March to May.

While myrtle likes to have a good

drink when dry, prolonged wet conditions cause entire branches to die. I have seen a bad attack of scale on a myrtle. If the plant is badly infected, root some clean cuttings and discard the plant.

Opuntia
(prickly pear cactus)

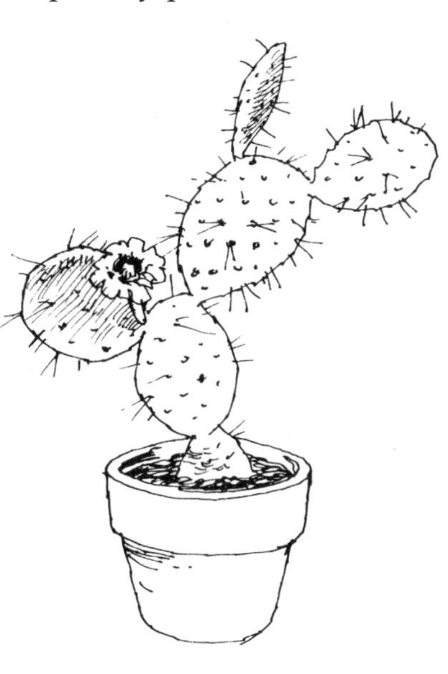

***Opuntia*—various species (prickly pear cactus)** Cactuses are not my favorite plants; I can get excited about them only when they bloom. However, I realize that many people get a great deal of enjoyment out of them, and there is no doubt that for hot, dry apartments, they are ideal plants.

Contrary to common belief, many poor cactuses die in cultivation from lack of water. Although they survive for months on end in the desert, conditions are quite different in an apartment or house. I water mine once a week, and even twice a week during the hottest part of the summer. It is also a myth that they bloom only once every seven years. Once they have started, flowering is an annual occurrence, with attractive 7-cm (3-inch) blooms that are either yellow or pink, depending on the species. With minimal fertilizing, open well-drained soil is the key to success. Don't allow direct hot sun from a south- or west-facing window to dehydrate them in the summer.

Oxalis purpurea
(flowering shamrock)

***Oxalis purpurea* (flowering shamrock)** The common name of this plant is misleading, as it is not a shamrock at all. It grows from a little bulbous tuber. In its native South Africa, it is considered a nuisance weed, yet in cultivation it is a very attractive plant. The trifoliate leaves are carried on individual stems about 15 cm (6 inches) long, as are the clusters of bright pink flowers, which usually keep appearing from February through June and July. In summer, when the plant begins to look sick, dry it off gradually for four to six weeks. Then shake off the old soil and replant the following January.

Flowering shamrock will do fairly well in poor soil, but small feedings are beneficial. Good light is a must, otherwise the plant gets too tall and falls over. If allowed, spider mites will really attack this plant, so keep a close watch and mist regularly.

***Pelargonium zonale* (pelargonium, geranium)** This is the common pelargonium on which so many of us rely for our summer show. I wouldn't particularly call it a houseplant, but it can be cultivated inside. Pelargoniums came originally from South Africa, where they reach shrub proportions. The first cross of the plant was produced in 1714. There have been so many since that now many of the 'Carefree' pelargoniums are grown annually from seed sown in January.

As houseplants, they need lots of light during the winter months. Artificial light prevents them from becoming straggly and keeps them blooming. All pelargoniums prefer poor soil and pot-bound conditions, which make them produce lots of flowers. If you've had pelargoniums on the balcony all summer and you want them to carry on indoors during winter, take cuttings in August, root them and grow them as more compact plants. Of course, you can always lift the whole plant and bring it in.

The pelargonium group is particularly worth investigating if you like plants with aromatic leaves. Various species produce rose, lemon or peppermint scents.

Pelargonium zonale
(pelargonium, geranium)

***Peperomia*—various species (radiator plant)** I don't know why this plant is known as radiator plant, except that it can tolerate low light, difficult conditions and drying out—perhaps even surviving on a shelf above a radiator! I like using them in terrariums, especially some of the smaller forms, such as *Peperomia caperata*, which is commonly known as emerald ripple because of its emerald, almost pleated,

Peperomia
(radiator plant)

heart-shaped leaves. Another favorite is *P. argyreia*, which also goes by the name of watermelon plant. Its waxy, heart-shaped leaves are marked just like watermelon skin.

As these plants are hard to kill, except by overwatering, they make fine housewarming gifts.

***Phaleanopsis ambilis* (moth orchid)**
This attractive plant is native to New Guinea. It will bloom almost continuously once established in your home, provided you can give it good humidity, air circulation and some artificial light. The growing medium is important with this one. Being epiphytic in nature, the plant grows on decomposing material caught in the branches of trees. Traditionally, orchids were grown in a medium that consisted mostly of osmunda fiber, but now most North Americans grow their orchids in a mixture of fir bark, moss and a little peat.

Phaleanopsis ambilis
(moth orchid)

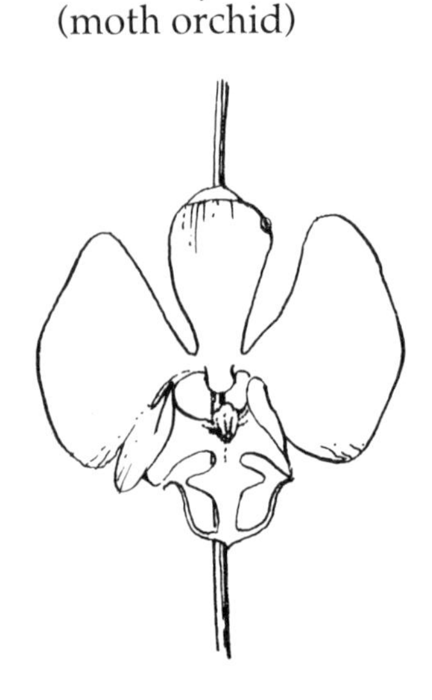

This particular plant produces about three or four basal leaves that are very wide, leathery and dark green. A long slender stem appears from the center. It can carry up to twenty blossoms, all about 2 to 4 cm (1 to 1 1/2 inches) long. Once the spike has finished blooming, it usually sends out another spike full of buds from the original stem, ensuring year-round blooming. The flowers are almost circular and white; however, there are other crosses that are yellow or even pink.

Philodendron scandens
(philodendron)

***Philodendron scandens* (philodendron)** This is the common small-leaved philodendron that can survive under all kinds of unusual conditions. I have even seen it growing happily in water over a kitchen sink, where someone had started to root it in water and left it because it was doing so well. I once had a friend who had a single-stemmed philodendron that was 15 meters (45 feet) long, with approximately fifteen leaves at the growing end!

Originally from Panama, this plant has gray-green or dark green leaves, depending on how much it is fertilized. It is a good plant for growing up the bare stem of a palm or something similar. It also looks very nice in a hanging planter if kept constantly pruned and bushy.

Cuttings consisting of a piece of the main stem and a leaf root easily. For example, the one that was 15 meters (45 feet) long would probably have yielded as many as twenty new plants had it been propagated.

Pilea cadierei
(aluminum plant)

***Pilea cadierei* (aluminum plant)** This was one of the first houseplants with which I fell in love. It is so easy to grow and doesn't seem to mind poorer light conditions—even to the point of surviving under a regular end-table lamp with the addition now and then of some early morning or late evening sun.

The leaves give it its common name. They are variegated, with a metallic sheen that contrasts well with its dark green background. It forms a bushy plant and responds well to drastic pruning back now and again, sending up plenty of strong new shoots. The pieces you cut off are very easy to root, so you can start a collection.

Another of this group, *Pilea microphylla* (artillery plant) is a good candidate for a terrarium. It is a small bushy plant with minute green leaves. If it is touched when the seeds are ripe, it shoots the seeds everywhere—the source of its common name. It can be a nuisance in a heated greenhouse, because it will seed itself in every available pot!

Pittosporum tobira variegata This plant grows throughout China and Japan, where it was discovered in 1804. It conforms well to the "balcony in the summer and inside for the winter" routine. It is quite a shrubby plant and in a 25-cm (10-inch) pot it will become about 1.3 meters (4 feet) square. The oval leaves are 2.5 cm (1 inch) long and a variegated cream white around the outer edges. During early summer, sweetly scented little clusters of cream flowers are bunched among them. There is also a plain green-leafed form.

Cuttings root well in May or June. Ordinary good potting soil is fine and a cool, well-lighted window is ideal during winter.

Pittosporum tobira variegata

Polianthes tuberosa **(tuber rose)** This plant is from Mexico, where it was discovered in 1629. During my grandmother's era, it was a popular subject for wedding bouquets because of the overpowering scent of its flowers. In the last couple of years, it has been appearing in many plant stores again.

The plant has a bulb-shaped tuber from which the common name is derived. The leaves are linear and grass-like, about 20 to 30 cm (8 to 12 inches) long; and the flower stem is 5 cm (2 inches) tall with large waxy white flowers at the top. Sometimes they are double and resemble a gardenia in a wedding bouquet. Its scent is also similar to that of a gardenia.

Tuber roses hate to be waterlogged and appreciate well-drained soil. I have

Polianthes tuberosa (tuber rose)

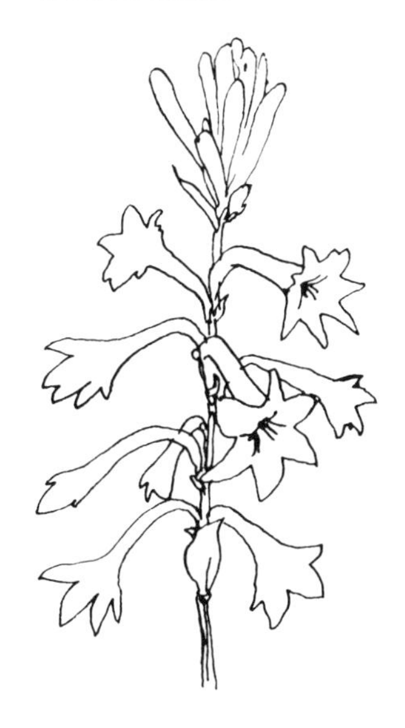

found that they seem to die out after about the second year in a pot. However, the flowers last quite a while, so tuber rose is well worth growing.

Polyscias balfouriana
(Balfour aralia)

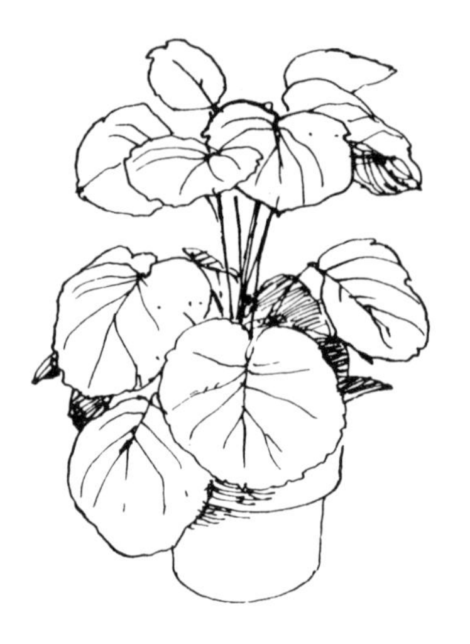

***Polyscias balfouriana* (Balfour aralia)**
This aralia seems to have gained popularity since a close relative, *Dizygotheca elegantissima* (false aralia), has done so well on the houseplant market. It requires similar treatment but it is bad for losing its leaves and picking up spider mites.

The stem of the plant is quite woody. The foliage is round, somewhat toothed, and marked similarly to pelargonium leaves with a cream white. Important factors are good light and a good soak when dry, but do not water it until it feels dry again. It is far too difficult to take cuttings, but air layering works reasonably well.

***Primula kewensis* (primula, primrose)**
This warm greenhouse hybrid primula grows well in a cool east-facing apartment window. The original hybrid did not set seed, but since its introduction in 1910, a seed-producing form has been developed.

The leaves are typically primrose form and the 15- to 20-cm (6- to 8-inch) flower stalks bear clusters of sweet-scented sulphur-yellow blossoms from January through April. Only occasionally is this plant grown commercially, even though it will go on for several seasons and can be easily propagated by division. The soil should be a rich mixture with lots of peat moss or similar moisture-retaining material.

Be warned though—a cousin of this plant, *Primula obconica*, does not get along well with some people. The leaf hairs can produce an itchy rash.

Primula kewensis
(primula, primrose)

Pteris cretica **(strap leaf fern)** This fern was a popular conservatory plant after its introduction in 1820. Its common name comes from the dark brown margins of spores on the undersides of the fronds. Carried on dark wiry stems, these fronds are 10 to 12 cm (4 to 5 inches) long and five-fingered. This plant is ideal for the bathroom or kitchen because it can tolerate low light intensity. Lots of direct light doesn't seem to bother it either. The soil should be moisture-retentive, yet well drained.

Pteris cretica
(strap leaf fern)

Rhoeo discolor **(boat lily)** This is an extremely easy houseplant to grow. It thrives weedily from Florida to Mex-

Rhoeo discolor
(boat lily)

ico. Its common name refers to its flowers, which are carried in a cup that is shaped like a boat, about 2 cm (3/4 inch) long. Boat lily has thick, pointed, succulent leaves that are dark green to purple in color. They all come from a central point and look somewhat like the top of a pineapple, with the boats tucked right near the base.

This plant requires moist soil at all times and tolerates poor light conditions. It spreads fairly rapidly and can be divided. Boat lily does well as a base plant in larger mixed tropical planters.

Saintpaulia
(African violet)

***Saintpaulia*—several species (African violets)** This is one of those lovely botanical names that is never used, as we know them all as African violets. This group of plants has a large following and I know people who have every available window ledge filled with African violets. A friend feeds hers with gelatin, getting the most incredible results and producing prize-winning blossoms.

They originally got their common name from early explorers in Africa who found them growing in damp spots on jungle floors and near waterfalls. The fleshy leaves are a lovely dark green and velvety. The blooms are violet blue. Of course, since their discovery, they have been hybridized, so that there are now pinks, whites, blues and mauves. There are even miniature forms.

For best success, they need humidity. However, they also need good air circulation to prevent fungus diseases. Water them by standing in a bowl of water for fifteen minutes, so that they can take the moisture from the bottom up.

Sansevieria trifasciata (snake plant)

***Sansevieria trifasciata* (snake plant)**
This plant, first discovered in tropical West Africa, is a great plant for those who neglect watering. However, it will respond well in moist soil and grow quite tall. It is often used in entrance planters in banks and office blocks.

The leaves are approximately 45 cm (18 inches) long and 5 cm (2 inches) wide. They are pointed at the top and horizontally marked with bands of smudgy light green and darker green all the way up. Very often larger pot-bound plants send up flower spikes which get everyone quite excited as the white-green blooms are arranged around the stem like a bottle brush. They open after dark and are highly scented. This plant loves poor soil but cannot stand being waterlogged. The

Scindapsus aureus
(pothos)

leaves may be chopped into 5-cm (2-inch) lengths for propagation. This is a good plant to start out with, as it is so hard to kill.

***Scindapsus aureus* (pothos)** This plant is originally from the Solomon Islands. It is very similar to *Philodendron scandens*, except that the heart-shaped leaves are variegated with yellow blotches and it is much slower growing. It does well in a hanging planter or an open terrarium, where it can be kept pruned back to a nice bushy little plant.

There is also a white variegated form that is even slower growing. A combination of stem and leaf cuttings works well, rooting easily at any time of the year. Many times pothos are sold with fiber stakes in the pot. They are great for the plant to cling to as long as the stake can be kept moist. I find a funnel of chicken wire filled with damp moss works much better.

Serissa foetida I do not know of a common name for this lovely little flowering shrub, which is sold quite widely. It is a native of Southeast Asia, where it was first collected in 1787. It has minute evergreen leaves that are less than 1.2 cm (1/2 inch) long. During late summer and fall, it is covered with double white flowers about .6 cm (1/4 inch) across.

I have found that this plant needs a fair amount of light. Even though the shrub grows slowly, it becomes pot-bound quite rapidly and needs to be potted on about twice each season. Fairly well-drained potting soil is necessary, with plenty of feeding. Cuttings taken during spring root easily.

There is also a slightly variegated form, which has leaves with edges that are thinly margined with a light gold color.

Serissa foetida

Solanum capsicastrum (Jerusalem cherry, Christmas cherry) Always available just prior to Christmas, this plant is a very easy plant for most people to grow as long as there is a balcony or patio. It is in the same family as the tomato but is not as pleasantly edible, so do not eat it.

If you are given one for Christmas, save all the orange berries as they drop off. When the outer case begins to dry out around March, break one open and sow the seeds in a pot of good sterilized soil in the kitchen. When they have germinated and the seedlings are large enough to handle, pot them singly. Then as soon as the dan-

Solanum capsicastrum (Jerusalem cherry, Christmas cherry)

ger of frost is over, put them out on the balcony. Pot on as they grow. The plants will flower around July and August. With luck they will attract bees and insects to pollinate them and many green berries will appear. At the first frost bring them back indoors, where the berries will ripen to a brilliant orange.

The main pests of this plant are aphids and spider mites.

Spathiphyllum wallissii (white flag)

***Spathiphyllum wallissii* (white flag)** A native of Colombia, white flag fares quite well in a woodland version of potting soil, which is two-thirds peat and leafmold. A friend has one that thrives and flowers in a north-facing apartment.

The 15-cm-long (6-inch) narrow leaves are glossy and the flowers are narrow and pure white. I've never had any insect pests on mine.

When white flag becomes crowded in its pot, division works well, especially during January or February, before too much new growth starts.

Stapelia hirsuta In the cactus and succulent group, this particular plant is interesting because it is so novel.

The branched stems are much like a thornless cactus. It is 15 to 20 cm (6 to 8 inches) tall, bearing star-shaped flowers, 5 cm (2 inches) wide that have a pale beige background with beautiful chocolate-mauve dots. The novelty is that these are carrion flowers, which look lovely but smell like rotten meat. The plant is pollinated by flies instead of bees! The scent can be detected only when the blooms are about 60 cm (2 feet) from your nose, so this isn't necessarily a plant to reject, as the odor won't affect the whole room.

Sun is important, but light should not be direct during the summer. Cuttings root easily in a good open soil. Like all other cactuses and succulents, it is susceptible to mealybug.

Stapelia hirsuta

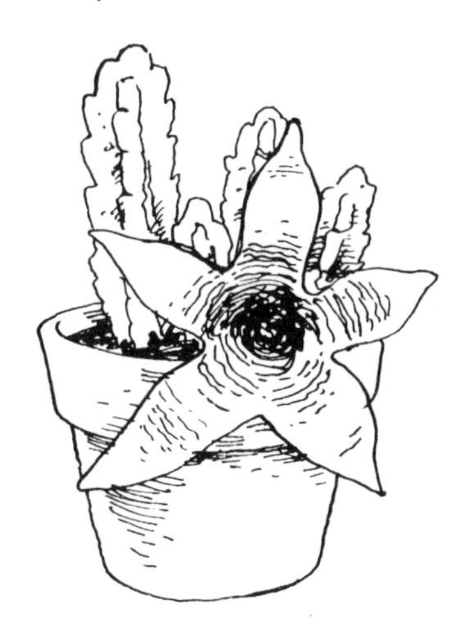

***Tolmiea menziesii* (piggyback plant)**
A native of North America, piggyback plant was first recorded by Archibald Menzies in 1812. It was later noted by Dr. Tolmie, who was with the first Hudson's Bay Trading Company settlement on the west coast of North America. Later it was named in their honor. In my home town, Vancouver, I have been told by some garden centers that it is a rare plant and in short supply. I have to smile, as it is native to this region and quite plentiful!

It makes a fantastic houseplant

Tolmiea menziesii
(piggyback plant)

for cooler rooms. A woodland type of soil, moist conditions and a semi-sunny window are essential. New plants can be started from leaves that already bear baby plants (hence its common name). Aphids are the main pest, but are not difficult to eradicate. Also keep a close watch for mildew.

Tradescantia fluminensis tricolor
(wandering spiderwort)

***Tradescantia fluminensis tricolor* (wandering spiderwort)** Many species of this plant are successfully cultivated. This particular species has leaves colored green, white and pink.

The important thing to remember is that the plant should be constantly cut back. There is nothing worse than a spiderwort high in a hanging basket with the only live tips 1 meter (3 feet) below it and all the rest brown and bare. Cut the tips off and root them in water or a mixture of peat and sand, then trim the original back to the rim of its pot. When the cuttings have rooted, put several in one pot to achieve a quick bushy plant.

Feeding of older plants is important, especially after the first three months. Regular potting soil is fine. If green shoots appear on variegated forms, remove them, otherwise they will take over the whole plant.

***Vallota speciosa* (Scarborough lily)** This South African native is a bulbous plant with strappy daffodil-like leaves and red flowers. It is much like a smaller version of amaryllis, but I find it more attractive. The bright scarlet flowers, which usually occur in the fall, are in clusters of three or so on a stem 60 cm (2 feet) long. The individual blossoms are 5 cm (2 inches) wide.

This plant seems to bloom better when completely pot-bound. Good sunlight is important, and it is interesting to note that Scarborough lily doesn't require a resting period. If you are fortunate enough to have a south-facing balcony or patio, keep the plant out all summer and don't forget to water it constantly. Well-drained poor soil is ideal for this plant.

Vallota speciosa
(Scarborough lily)

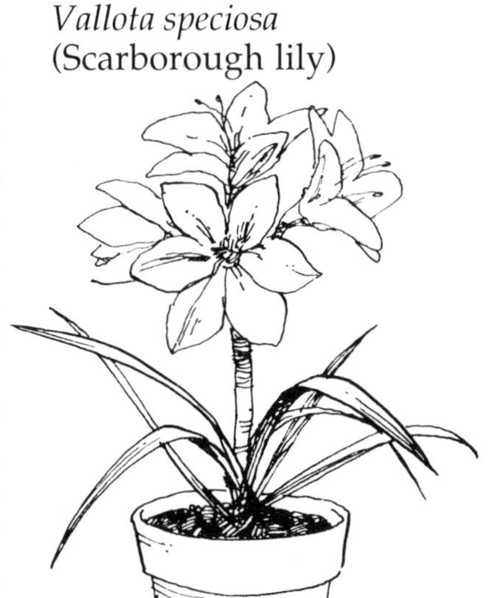

Veltheimia capensis
(forest lily)

***Veltheimia capensis* (forest lily)** This is another South African native that is not grown as widely as it might be. It is a fairly big, bulbous plant with large leaves up to 5 cm (2 inches) across that are shiny green and somewhat crinkly along the edges. The thick flower stalks appear in winter and are about 45 cm (18 inches) high. All around the top, delightful bright pink tubular flowers occur. They don't last too long but are plentiful during the blooming period, which lasts four to six weeks. If the

Zygocactus truncatus
(Christmas cactus)

flowers are pollinated, large three-cornered seed pods occur, which are white when they eventually dry. The black seeds inside germinate easily if sown right away.

This plant remains green year-round. It needs virtually the same treatment as the preceding plant. If not checked closely, mealybug can become a problem.

***Zygocactus truncatus* (Christmas cactus)** A native of Brazil, this plant bears orchid-pink flowers. It is a jungle cactus, as opposed to a desert cactus, so it needs a much richer soil with plenty of moisture-retaining material. It also requires regular feeding, especially leading up to flower time. You should also be warned—this plant can cause heartache. A lot of people have trouble getting it to bloom.

Christmas cactus hates to be repotted. It loves cramped roots. A cool window sill away from drafts, and around 17°C (63°F) is a good position. Light is also an important factor. Like the poinsettia, it requires twelve hours of darkness each night for a couple of months before Christmas. It hates to dry out when in bud or blossom and does not like to be moved around. However, a six- to eight-week dry period in March and April is beneficial. The individual pads or stem segments propagate easily if taken in spring. Always try to have a younger plant to hold onto in case the old one doesn't survive.

IV

Vegetables and Herbs

When it comes to growing plants, don't think you have to limit yourself to houseplants and ornamentals just because you live in an apartment. Quite a few vegetables can be grown in containers. I didn't believe it possible until I tried it myself. The thing to remember about growing vegetables this way is that essentially the soil mix is the same as for ornamentals on a balcony. The major difference is that you should not be tempted to overcrowd vegetables in limited root spaces. For example, to grow tomatoes successfully, one plant per 25-cm (10-inch) pot is the answer.

Many people are using no-soil mixes made up of peat, vermiculite and sometimes perlite, all of which have no food value whatsoever. But all the basic and essential nutrients can be added—rather like hydroponic growing. Getting the right balance of nutrients is quite a tricky business and deficiencies can be disastrous. One homemade mix you can use consists of one part moist peat, one part vermiculite and one part perlite. To each 12-liter (3-gallon) pailful, add 15 ml (3 tablespoons) of a slow-release, high phosphorous 6-8-6 fertilizer and 5 ml (1 tablespoon) of lime. Mix thoroughly.

Those three numbers that always occur on bags of any fertilizer are important to any growing plant. The first number, which stands for nitrogen (N) is for leaf and stem development. Phosphate (P) is for healthy root growth, and potash (K) is a general balancer, giving flowers and fruit. The 20-20-20 fertilizer I mentioned earlier for flowers and houseplants provides equal amounts of the three elements. If you are growing leaf crops, such as lettuce, spinach or cabbage, 10-6-4 is a good fertilizer; for root vegetables use 6-8-6; and for fruit, 5-10-10. The elements are always in the same order—nitrogen-phosphate-potash.

Container Vegetables

The following is a list of vegetables that can be grown in containers.

Beans are fun to grow. 'Scarlet Runner' is ideal for growing screens, giving both colorful blossoms and delicious beans. To get an early start, germinate them and plant them in small pots about three weeks before putting them outside. The time to put them out will vary according to where you live, but usually it is the last week of May or the first week of June. Beans cannot stand frost.

Three or four plants are ample. They will take over by the end of the summer if you let them! In my particular area last spring, 'Scarlet Runner' seeds were hard to get. Everyone wanted to grow them. Because so few plants are needed, all you require to start with are six beans. Let a few pods mature and dry so you won't have to buy them next season.

If you do start the beans indoors, soak them for a couple

of hours in a cup of water. By then they will have swollen quite a bit. Drain off the water and put the beans between a few layers of wet newspaper for two days until they have sprouted. Then plant them in individual 7-cm (3-inch) pots, covering them with about 1 cm (1/2 inch) of soil. Be careful not to break the sprout. Before putting them on the balcony, plant about three to a 25- or 37-cm (10- or 15-inch) pot. Later, when they start to get out of hand, you can trim them with no danger to the plant.

If you don't have room for runners, start bush beans the same way, with three per pot. Two good varieties are 'Tendercrop' and 'Pencil Pod'. All beans prefer sunny locations.

Beets can be grown well in containers. Of course, the young foliage is as tasty as spinach. Beets are sown directly into large pots or boxes about 22 by 45 cm (9 by 18 inches). Scatter-sow thinly; later you will need at least 10 or 12 cm (4 or 5 inches) between each plant. Chances are you will harvest them while they are still young. 'Early Wonder' and 'Ruby Queen' are two good varieties. Semi-sunny balconies are ideal for beets.

Brussels sprouts are usually planted four per apple crate or 36-liter (1 bushel) container, with one in each corner. Since the plants tend to become rather top-heavy, some kind of sturdy stake .6 to 1 meter (2 to 3 feet) long will be required. Cabbage root fly can be a problem in some areas; however, coffee grounds scattered among the plants, or reemay cloth work well as a deterrent.

Brussels sprouts like lots of feed that is fairly high in nitrogen and they need at least four to six hours of sun each day to grow successfully. They like cooler temperatures, so in hot areas, place them in an east or north exposure. 'Jade Cross Hybrid' and 'Long Island Improved' are two early bearers.

Carrots need deep pots and only short-horn varieties should be chosen for containers. Scatter-sow seed thinly.

Later, when the plants have to be thinned, carrot root fly may attack. I'm not a great believer in many of these old remedies; however, if carrot fly is a problem in your area, coffee grounds sprinkled among your carrots on a weekly basis will keep the fly away. You can also cover the container with reemay cloth, but keeping it from blowing away could be a problem.

Don't be tempted to pull up the carrots too soon. But if you do, use the foliage in a bouquet with cut flowers that you have grown on your balcony. Carrot greenery always looks nice with sweet peas. Two good short carrots are 'Oxheart' and 'Royal Chantenay'. All carrots will tolerate both sun and shade.

Cucumber plants look nice if they are trained up a trellis in a sunny, sheltered area of the balcony. The vines can be pinched off after about 1 meter (3 feet). These produce sideshoots that bear cucumbers.

Cucumbers (and zucchinis, squash and others in the same group) often show masses of male flowers before the females, so don't lose heart when enormous blossoms fail to produce anything. When you do get two blooming together, try to pollinate the females, which are the ones with the bump directly underneath the base of the flower.

Always harvest cucumbers regularly to ensure a continued crop. Sun is essential. If powdery mildew is a problem in your area, plant them in good open places where the air circulates freely. 'Burpees Hybrid' is a very successful variety. For small cucumbers, try 'Patio Pik'.

Green onions can be grown successfully in a 30-cm (12-inch) clay pot. Scatter-sow the seeds as soon as the weather is warm enough. If it is a good season, often a second crop can be sown in the same pot after the first has been harvested. 'Evergreen Bunching' or any standard variety of onion can be grown for bunching.

Onions do not attract aphids unless they are grown among other plants infected with these pests. Green bunch onions can tolerate sun or shade.

Lettuce is not one of the easiest crops to grow in containers. I have never had success with any of the hearting varieties. But 'Grand Rapids' leaf lettuce is ideal and will do well in shade because it likes quite a bit of moisture. This variety never hearts but it produces many large succulent leaves, which can be picked frequently from the outside. More will be produced all the time. Often the plants will last through the summer, but if they don't, you can grow more. Lettuce is a quick, easy crop.

Peas are so delicious when picked and eaten fresh. They need soil at least 30 cm (12 inches) deep in a box approximately 1 meter (3 feet) long and 22 cm (9 inches) wide. Peas like a lot of humus and moisture in the soil. I have grown 'Little Marvel' in a container, as this variety is only about 45 cm (18 inches) high and is easy to cope with in a limited space. A friend tried 'Snow Peas', which grow quite tall. Only a few plants of this variety are needed, since each one produces many peas. Germinate peas the same way as beans, using layers of newspaper.

Pepper plants are attractive but can be grown only on hot sunny balconies. For balcony gardening the smaller chile varieties are probably best. Certainly they are the most colorful. 'Red Chile' is very hot; 'Stokes Early Hybrid' is a good variety for Canada. Poor setting occurs when the temperature drops lower than 16°C (62°F) at night. But hand pollinating with an artist's brush works well if done midday. Just dust the brush from open flower to open flower. Peppers take a lot of water. A 22-cm (9-inch) pot is ideal for each plant.

Radishes flourish in balcony containers and can be sown all summer. Scatter-sow them in a 30-cm (12-inch) pot. If you keep them well watered they should be ready for harvesting in a month. Three to four hours of sun a day is essential.

Squash plants are very prolific—one is definitely enough. If you are worried that it will take over the whole balcony, grow one of the bush types. Last summer I had one in an 8-liter (2-gallon) plastic nursery container. I pinched the top out when it was young, with the result that I got two shoots that ran 2.5 meters (8 feet) in opposite directions.

Squash and its relatives love rich compost in a garden and are often grown on the top of compost heaps. That is an indication of what soil you should use in a container. 'Early White Bush Scallop' is a good variety for balconies. Squash and zucchini are pollinated the same way as cucumber. Half a day's sun is essential.

Swiss chard is a must if you like spinach, because the leaves are eaten in exactly the same way. The bonus with Swiss chard is that the thick white stems can be steamed separately and served with melted butter. They taste somewhat like asparagus. The seeds should be sown as soon as the weather warms up, so that you end up with four plants per 25-cm (10-inch) pot.

If you are especially fond of Swiss chard, plant more than one pot. This is the sort of vegetable that grows all summer and will continue producing in mild winters. After you cut it, it comes again. Two good varieties are 'Fordhook Giant' and 'Rhubarb Chard'. This vegetable grows well in shade.

Tomatoes have probably been tried on most balcony gardens. Sun is the most important factor with these fellows. The secret is to try not to get too much from one plant. It is best to purchase the plants and place only one in each pot.

Regular tomatoes need to be well staked. Train them up by pinching out the sideshoots as they appear. Try to get the first bunch of flowers to set well. Sometimes this is quite a tricky business. The best way to pollinate tomatoes is to tap the whole vine around mid-day, when the pollen is nice and dry, so that it flies around in the air, thus setting fruit.

If you get four bunches of fruit from one plant, take the growing top out and remove some of the basal leaves. This way, all energy will go into producing tomatoes rather than growing the plant taller. Some of the bush and so-called patio tomatoes constantly need support to prevent the plants bending as the fruit sets. With tomatoes in pots it is extremely important to keep them moist at the root at all times while the fruit is developing. If you don't, you will get blossom end rot, when the base of the fruit dries out and goes black.

Some good varieties of tomatoes for pots are 'Sweet 100s', which grow quite tall but produce masses of tasty cherry tomatoes, and 'Pik Red', which is a very abundant producer in a pot.

Admittedly, none of the vegetables I have mentioned will yield vast quantities, but they will give you the satisfaction of eating something you have grown yourself.

Growing Herbs

If you haven't room for vegetables, a pot or two of herbs or a herb jar (alias strawberry jar) is a must. The latter has become quite popular over the last few years and is very attractive, especially if a few flowers are mixed in with the herbs.

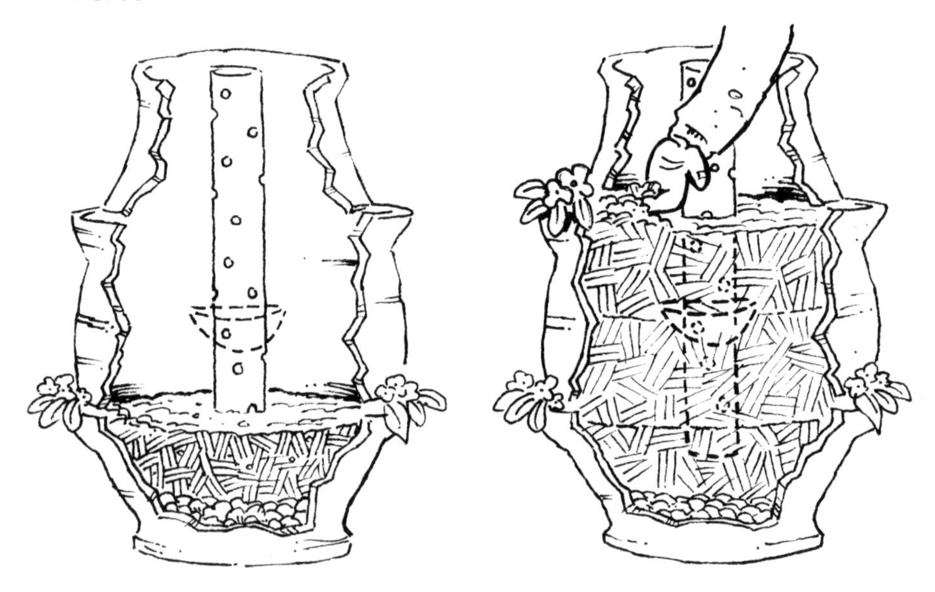

Herb pots come in various sizes. They are usually made out of pottery, so before using, soak them in the bathtub. If possible, try to get a pot with a hole in the bottom.

Put a layer of broken crockery or some pebbles in the bottom and fill with sterilized potting soil, stopping when the first layer of holes is reached. Firm it finger tight, then lay in the first three herbs, perhaps parsley, chives and oregano, with the growing tops completely out of the holes. Put some tubing, such as a piece of 10-cm (4-inch) downpipe, 30 to 45 cm (12 to 18 inches) long, down the center of the pot so that it is hidden inside and just level with the rim. Then fill in around the pipe up to the next layer. The tube ensures that the plants at the bottom get enough water. Without the tube, water would run out of the top holes without reaching the bottom plants. Plant two more herbs, perhaps thyme and sage, and one trailing lobelia. Fill to the neck of the container or about 1 cm (1/2 inch) below the rim. Plant chives, parsley and a gazania that will cover the top and hide the hole.

In very cold areas, herb gardens are good only for the summer, but in more temperate areas they can be left out all winter. Of course, it is easier to buy small herb plants, but many of them do quite well from seed and can be started off indoors. Do not sow the seeds too early or they will damp off in the heat of the apartment.

Mint should never be planted in a herb pot, as it takes over and grows out of every hole. If you are fond of mint, grow it in a 25-cm (10-inch) clay pot. It is a most attractive plant among flowers on a balcony. Mint grows readily from cuttings, which may be started in water, although peat and sand is better.

Herbs such as parsley and chives tend to flower as they age, and I always let them bloom since they are so pretty. When the flowers start to fade, I cut them off and I also cut half the plant down to the base. I do the same with mint to encourage new growth. When it gets going, I cut down the other half. In this way, I always have tender young growth to eat.

Here is a short list of easy and useful herbs to grow from seed:

Sweet basil	sauces
Caraway	seeds add zip to cakes and breads
Chives	excellent in salads
Dill	fish dishes
Marjoram	delicious with mushrooms and zucchini
Mint	for lamb dishes, barbecue and salads
Oregano	tasty on vegetables, in salads and meat

Parsley	parsley sauce, garnishes and salads
Sage	stuffing in pork and fowl
Tarragon	used in various dishes, especially as vinegar flavoring

There are, of course, many more. While most herbs need a full sun location, parsley, mint, sage and thyme will tolerate a north-facing balcony.

Growing Herbs and Vegetables with Hydroponics under Lights

The term "hydroponic" means growing plants without soil. Instead, they are grown in water, to which the necessary nutrients and micro-nutrients are added on a regular basis. I know it sounds terribly scientific and complicated, but it really isn't. In most large cities throughout Canada and the United States, there are companies and garden shops that carry home hydroponic units suitable for apartments and townhomes. If you live near a center that specializes in this field, you will be able to purchase all the supplies from them.

If you have a balcony, you may only need to use the unit during the winter months, to keep yourself supplied with herbs and greens. If you have the space, you may also want to keep it going outside on your balcony during the summer.

A couple of tips about using this system inside during the winter are in order. First of all, you will need the addition of some artificial light to simulate the daylength we get during the summer months. Most of the units sold come with the light fixture added. If not, you will be able to purchase it from the same place. If you are setting up your hydroponic garden next to a window where you will get all the available winter light, you will only need to use regular fluorescent tubes to supplement the available daylight. A combination of one cool white and one warm white will be fine.

On the other hand, if you are growing in a poorly lit area away from a window, you will need to use special growing lights that simulate as closely as possible the full spectrum of natural sunlight. Don't even think of using a sunlamp. It will burn your plants and elongate them beyond recognition.

Whichever form of lighting is used, it will need to be attached to a timer switch so that it is on at least twelve hours every day. Fourteen hours is even better.

If the light fixture is not attached to your growing unit, you will need to suspend it with chains or something similar, to allow you to adjust the height to suit your crop. The light source should be about 30 to 45 cm (12 to 18 inches) above the crop for best results.

Temperature is the other key factor in this type of growing. If you live in an apartment block where you can't regulate your own heat, you may run into some problems. A daytime temperature of 23°C (74°F) is fine, and a drop to 18°C (65°F) at night when the lights are off would be preferable. This then simulates nature.

As you might imagine, all this costs a few dollars to operate. But it is fun to be able to grow your own leaf lettuce all winter long, along with parsley, chives, basil and many other herbs.

Another nice thing about it is that in the drier areas of the country, a hydroponic unit can help keep the humidity at a better level in your home environment.

V

Terrariums and Bottle Gardens

Bottle gardens and terrariums are ideal ways of having some houseplants in your apartment. This is especially true if you are one of those people who forget to water plants, or if you have to be away a lot. A friend of mine moved from Ontario to British Columbia and her terrarium survived two weeks in the moving truck and storage.

In many old horticultural books, terrariums are referred to as Wardian Cases, so-named for Nathaniel Ward, a London physician who, while studying entomology, noticed moss and a fern growing in one of his tightly sealed specimen jars. He was so fascinated that he switched his field of research to mosses and ferns. Later he maintained that they would live up to 200 years in properly enclosed conditions. Of course, no one has lived long enough to test that theory; however, I have had a large bottle garden growing happily for three to five years.

The plants you can use in gardens of this type are rather limited and you will have to experiment. The following is a list of plants that do well in terrariums.

Calathea makoyana (peacock plant) This is a very close relative of the prayer plant, but it has a much more compact habit. The leaves, which also close at night, have beautiful dark red blotches, which give the effect of peacock feathers. It propagates easily from division. This is one of the plants in my bottle garden.

Cryptanthus bivittatus minor (earth star) A member of the Bromeliad family, it looks just like a little starfish with pink foliage and a darker pink area down the center of each leaf. Earth star roots itself so easily that it can be dropped into a terrarium without any roots. When happily established, it produces offshoots from the center of the plant.

Dracaena surculosa (gold spots) This plant can be a bit large for terrariums, but it is ideal in its younger stages and can be pruned back in a bottle garden. The stems are cane-like, with long, dark, leathery leaves 4 cm (1 1/2 inches) long at each joint. The foliage has yellow blotches all over the surface. Propagate by cuttings.

Fittonia Verschaffeltii (nerve plant) This low-growing plant is a tropical groundcover. It has oval leaves about 4 cm (1 1/2 inches) long that are a very showy dark green with white veins. There is also a pink-veined form *F. argyroneura*. These are some of the most successful bottle garden plants that I know. They root easily by layering the stems, a process that occurs naturally on a healthy plant.

Maranta leuconeura (prayer plant) This plant will fill a bottle garden in six months, so it is ideal for a fast effect. The only trouble is that it tends to overcrowd other small plants that might be in the bottle. I had one in a terrarium and I pruned it back three times before I finally had to remove it.

Peperomia—variety of species There are many species of this plant, and all are extremely good terrarium plants.

They have thick glossy leaves that are almost waxy and could be confused with succulents. However, the common factor that identifies them is the white or gray mousetail-like flowers that grow up to 15 cm (6 inches) long.

Pilea cadierei (aluminum plant) This is a large terrarium plant, reaching 22 to 30 cm (9 to 12 inches) in height. It is a fairly bushy plant with pointed oval leaves 2 cm (1 inch) or so in length. Two broad metallic splashes on each leaf give it its common name. The aluminum plant propagates readily from cuttings and can be pruned back in a bottle garden, only to shoot out again.

Pilea microphylla (artillery plant) Reaching a height of about 12 cm (5 inches), this is a very branched plant with hundreds of minute light green leaves that give it a fern-like appearance. Wherever the leaf joints occur there are even tinier flowers which, when touched, release clouds of pollen. This plant readily seeds itself and will root easily from cuttings.

Selaginella kraussiana (spreading club moss) This is a must for any bottle garden because, of all the plants, it is the most at home in a terrarium. Moss-like and low-growing, it produces leaves arranged something like a fern. Spreading club moss used to be impossible to buy. One would have to visit a large municipal or commercial greenhouse, drop something, and while retrieving it, accidentally pick up a piece of club moss at the same time! It grows rampantly under greenhouse benches.

You will notice that all of the plants mentioned are foliage plants with only insignificant flowers. Varieties with large flowers, such as African violets, would love to grow in a bottle garden. But in a completely sealed environment, the dying blossoms would create such humid conditions that the whole garden would start to rot, even if all the plants were very healthy. However, blossoming plants are excellent in an unused aquarium with a light cover that can be lifted off for the removal of old leaves and blossoms.

I did not mention ferns in my list because their fine leaves rot when they come in contact with glass walls. When I use them in a terrarium or large bottle, I always put them right in the middle so that it takes a while for the fronds to touch the edge. Many people ask me why I don't recommend native woodland plants and mosses. The reason is that the heat of our apartments is too much for cool-growing North American plants.

All of the plants in the list can tolerate low light conditions as they are native to jungle floors which get no direct sunlight. My most successful bottle garden has a cork lamp fixture in the top with a grow-light bulb and a dark lamp shade. It makes a fascinating end table lamp and is a great idea for a gift.

Making a Bottle Garden

To make your first bottle garden, try a 2-liter (1/2-gallon) or 4-liter (1-gallon) wine jug. If you don't have one of these, try an old pickle jar with a screw top, a fish bowl, an apothecary jar or a brandy snifter. Whatever container is used, make sure it is well washed. Clear glass is preferable to green. If the latter is all you can find, then you will have to use plants with light variegated leaves so they will be visible.

Beware the many large ornamental bottles produced especially for the terrarium rush. Many of them are made of very thin glass and the results can be disastrous. If you use a bottle that was originally made to hold liquid, then you know it is a strong glass. That is why a wine jug is ideal.

Make sure the neck is large enough to get a cork, teaspoon and a fork through without difficulty. These three are essential for planting and a fork is handy for removing unwanted and dead plants. All the tools should be securely fastened to canes or rods with masking tape and the cork

could even be glued for extra security. Otherwise you might end up with a variety of corks and spoons in the garden.

The plants will be in small pots, so tap them out and wash the soil off their roots. This can be done ahead of time and the plants kept in a plastic bag until you are ready. The soil mix I use consists of one part sterilized soil, one part moist peat and one part coarse sand, which have all been passed through a .6-cm (1/4-inch) screen and well mixed with one handful of crushed charcoal per 4 liters (1 gallon) of mixture. I do not add any fertilizer, because the plants would grow far too rapidly. The mixture should be so moist that it sticks together when squeezed. If you are unable to make up your own mix, buy prepackaged African violet soil mix.

The other things needed are a small amount of well-washed pea gravel, a piece of cotton wool on the end of some coat hanger wire for cleaning the inside of the glass, and a funnel made out of an old bleach bottle with the bottom cut out.

Using the funnel, put about 1 cm (1/2 inch) pea gravel on the bottom, followed by 4 to 5 cm (1 1/2 to 2 inches) of soil.

If you have a large bottle you need 10 cm (4 inches) of soil. Because it is moist it is difficult to get through the funnel, so use a cane. The soil inevitably ends up as a mountain, which is then leveled out and tapped down with the cork. Do be careful not to put in too much soil. This can happen easily when filling a bottle from above as you get a distorted view through the glass. Obviously the soil doesn't end up as firm as garden soil but this is not important.

When the soil is leveled, tilt the bottle slightly to one side and scoop out a hole with the teaspoon. If you dig up some of the gravel it's all right. Then take out the spoon and drop in a plant, aiming for the hole. If there are a lot of leaves, you will be quite surprised at how small you can fold them up to get them through the neck.

Push the plant upright with the cork and firm the soil around it. If you cannot get it completely straight, don't worry. After a few days the plant will straighten itself. Put the other plants in the same way, by tilting the bottle. With a large bottle you could put a plant directly in the center. Three plants are ample, even if they do look lost at first. After a couple of weeks you will be amazed at the rapid growth.

After all this the inside of the neck is bound to be dirty, so lightly mist around the neck. But be careful not to spray too much moisture, because of the lack of drainage. I find the best way to clean the inside is with the cotton wool on a wire. It can be bent to reach all areas, rinsed off and used over again. When you have finished, stand the bottle in a shady area of the room. Very likely the next day it will be full of condensation, which will decrease over the next seven to ten days. When there is hardly any condensation left, screw on the top. Should one of the plants die, remove it by sticking a fork into the center to twist and pull. If a plant gets too big and needs pruning, securely fasten a sharp knife on a cane and saw away at the stem. After it is cut off, remove it with a fork.

When the whole bottle becomes overgrown, half fill it

with water and empty it. Repeat the procedure until all the soil is washed out. Then pull out the plants and roots with a bent coat hanger. This ruins the plants, but parts of them can be salvaged for cuttings.

Closed bottle gardens never need watering or opening, since the plants give off carbon dioxide at night and oxygen during the day, and the moisture is recycled. With regard to feeding, obviously a bottle that has been growing a year or more needs some food, so just add a few drops of some diluted plant food about every eight months. If you make an open terrarium, it's an easy matter to save the plants and pot them up. Cut down on watering with an open terrarium by fitting a piece of glass over the top.

Hanging Bottles

A variation on the terrarium is to grow a plant from the base of a wine bottle. This is done primarily to hang outside for the summer, although I have seen houseplants grown in the same way. The diagram on this page will give a better idea of the effect that can be achieved. Only dark wine bottles with deep dents in the base should be used. Since I always think that leaving the label on looks rather good, I spray a few applications of varnish to help protect it.

The easiest way to get the hole in the bottom is to use a hammer and a 15-cm (6-inch) spike. Lay the bottle on its side on the floor and hold it securely with the left foot. Then with the hammer and spike, give one sharp tap in the center of the dent. This usually makes a hole 1 to 1.5 cm (1/2 to 1 inch) across. Sometimes splinters of glass fly out through the neck, so it's a good idea to leave the cork in. If some of the broken glass stays inside the bottle, don't worry about it.

Remember that the dented bottom is very important. Some people neatly cut out the whole bottom of the bottle. This is great when planting, but when the bottle is hung

upside down the entire contents fall out! When you have successfully made a hole in the bottom, put either some moss or foam rubber in the neck, turn the whole bottle upside down and prop it up with newspaper padding in a flower pot or something similar.

Fill the inside of the bottle with sterilized potting soil and try not to be tempted to push it through the jagged hole with your fingers. Firm the soil with a stick as you go. Obviously one can't get the soil up under the ridge around the hole. But later, when the bottle is hung right side up, it will fill in gradually. As long as the soil is firm up to the hole, all will be well.

Make a hole in the soil with a pencil or a stick and insert a young bushy plant. In order to fit the plant through the hole, you will have to wash the soil off the roots. For obvious reasons, try not to press the stem of the plant too hard against the jagged edge.

After planting, stand the entire bottle and pot on a bright window sill, watering about every five days. After three weeks, roots should be visible through the dark glass. This indicates that the bottle can be hung upside down.

Twist fairly heavy wire around the ridge at the neck and make a hook for hanging it. Remove the moss or sponge from inside the neck and water it from that end.

It is essential that bushy plants be used. Some good ones are: impatiens, particularly the 'Elfin' strain; miniature pelargoniums; and multiflora petunias, which are the small-flowered varieties. Whatever plant you use, remember to pinch the top out after it has about six leaves, so that there will be six branches all around the bottle. Don't ever be tempted to use trailing plants, as they hang straight down and look terribly uninteresting. I have made gifts for friends using bottles with either parsley or sage.

VI

Gardening for Children

Apartment children above all others need to be introduced to the excitement and wonder of watching something grow. Since quick results are the key to capturing a child's interest, one of the best plants to start with is curled cress. The seed is available in any garden shop. All you need is a roasting pan or any dish about 5 cm (2 inches) deep. Place three or four paper towels in the bottom. Thoroughly moisten this pad of towels, then sow a layer of cress seeds approximately .6 to 1 cm (1/4 to 1/2 inch) apart.

Cover the container with newspaper or something that will completely cut out the light, and stand it on the kitchen counter or table where it can be easily inspected each day. Depending on the time of year the seed is sown, it will take from one to three days to germinate. Then remove the paper and allow light (not direct sunlight) to fall on the tiny plants.

On the fourth day, when the plants are 5 cm (2 inches) high, harvest them with scissors and use the tasty cress in salads or sandwiches. I have seen the seeds sown in a name or shape, which is even more fun. Cress can be sown at any time of the year.

Another plant that your child can grow quickly is the bean. Just imagine the satisfaction children would get if they actually sowed the beans you ate in the summer. Earlier in this book, I mentioned soaking the beans before sowing, using the wet newspaper method to start them. Your child can do this, but when the beans have started to sprout, line a tumbler or peanut butter jar with paper toweling or blotting paper, then fill with sand or moist peat. Slide the germinated beans down between the paper and the glass, four per jar. The peat or sand should be kept very moist. Most important, the jar should not have a ridge at the top, so that the bean plants can be tipped out at a later stage when they are ready to plant. The idea is to enable children to watch the roots and the top grow from the seed.

This project should be started about three weeks before the beans are to be placed on the balcony. Of course, planting time varies according to which area of the country you live in. The beans continue to be interesting outside, particularly because they climb counterclockwise. As a child, I tried twisting them clockwise, only to find the following day that they had reversed themselves.

Here is another project that is fun for children and it can be done inside or out. If it is an inside project it will be short-term; on a balcony it can last all summer. First paint some arms and buttons on a 12- to 15-cm pot. Then fill it with drainage over the hole and some sterilized soil. Moisten the soil and sow grass seed on the top. Water it well and cover with newspaper until it has germinated. Hairstyles created with the grass can be changed all summer long. Watch for drying out, though, as the head pot is only 10 cm (4 inches) across. If your children like this idea, you can experiment with many types of plants for the hair.

A great way for children to grow bulbs is in water. We used to be able to buy nice hyacinth growing jars that were shaped like egg cups, but any clear, narrow-necked glass jar works well as long as a bulb can be placed on top without falling in. Instead of putting this outside in a cool place, it is usually put in a cupboard. But the dark place has to be cool, and if you don't have a cool cupboard, then use a cardboard carton that can be closed. This method works quite well on the balcony, because bulbs need 5°C (41°F) for six to eight weeks.

If your children are particularly interested in this idea, try growing paperwhite narcissus as well. Pebbles are used to wedge the bulbs upright in a bowl 7 cm (3 inches) deep. Keep the water level up at all times and observe the cool dark period.

In both of these cases, specially prepared bulbs are used. They should be started from mid-September to the end of October.

Another old idea that delights children, especially in the depths of winter, is to save the carrot tops that are cut off and usually thrown away. Just stand them in a shallow dish of water, where they will quickly sprout green leaves. This is a short-term project that rarely fails. Actually, any tops off root vegetables will work.

For older children, an easy method of topiary can be achieved by twisting chicken wire into different shapes, such as a rabbit or bird, and filling with moss. Firmly anchor the shape in a flower pot. Plant an ivy on each side, which is then trained up over the wire. Keep the moss moist at all times. As the ivy grows, it can be pinned down with little pieces of bent florist wires. Attempt this only with children who have patience, as ivy is rather a slow grower. However, once it gets a hold, it will grow quite rapidly and will have to be trimmed to shape. Another plant that can be used in this way is *Ficus pumila* (creeping fig).

Many interesting plants can be grown from seeds of fruits that are part of our everyday diets. Avocado is grown by many adults, with or without success. By the way, there is no need to stick toothpicks in an avocado seed. I just soak the seed overnight, then plant it in a 10-cm (4-inch) pot of sterilized potting soil with about 1 cm (1/2 inch) of soil over top. An avocado won't germinate if the fruit was harvested before it was properly matured. When it has grown to about 30 cm (12 inches) and has about six leaves, cut the top out so that a branch will grow from each leaf joint and make a bushy plant.

Other seeds that grow well this way are grapefruit, lemon and orange. Grapefruit seems to be the easiest and very often during the summer the seeds have already started to germinate in the fruit. All of the citrus fruits can be put outside in a sunny position on a balcony in June, July and August. With the exception of lemons, it is quite unusual to get any citrus grown from seed to fruit. Other fruit seeds that germinate readily are grapes and apples.

Sweet potato and yams grow well in pebbles and water, and will climb all around the window.

Index